FEE MINING AND MINERAL ADVENTURES IN THE EASTERN U.S.

By

James Martin Monaco
&
Jeannette Hathaway Monaco

Published by:
Gem Guides Book Co.
315 Cloverleaf Drive, Suite F
Baldwin Park, CA 91706

Copyright © 2004
Gem Guides Book Co.

First Edition

Printed and bound in the United States of America

Cover Design: Scott Roberts

Library of Congress Control Number: 2003102004
ISBN: 1-889786-27-6

DISCLAIMER:
Due to the possibility of personal error, typographical error, misinterpretation of information and the many changes due to man or nature, *Fee Mining and Mineral Adventures in the Eastern U.S.*, its publisher and all other persons directly or indirectly associated with this publication assume no responsibility for accidents, injury or any losses by individuals or groups using this publication.

In rough terrain and hazardous areas all persons are advised to be aware of possible changes due to man or nature that occur at the various collecting sites.

"There is something about treasure that fastens itself upon a man's mind. He will pray and blaspheme and still persevere, and will curse the day he heard of it, and will let his last hour come upon him unawares, still believing that he missed it by only a foot."

– Joseph Conrad –

Good luck hunting for your treasure. May this book be a resource to aid your search.

– Jenny & Jim Monaco –

MISSOURI

NEW HAMPSHIRE

NEW JERSEY

WISCONSIN

We have always enjoyed traveling and treasure hunting. We have panned for gold, dug for precious stones and searched beaches for Spanish treasure—meeting great people everywhere we go. However, there have been many times after a trip when we discovered that we were within minutes of a place to mine and unaware it existed. It frustrated us to no end to miss opportunities to enjoy our hobby. Another frustration was to travel to a gold-bearing area of California, only to find that every square inch of territory was covered with mining claim posters. We decided to search for a source book that would help us find places to treasure hunt and rock collect. To date, we have not found that book. So, we did the next best thing—we wrote one ourselves. Now you and your family won't miss a collecting opportunity that is only a few miles away, just because you never knew of its existence.

There are over over 75 digging sites in this book where you can try your luck at finding treasures as varied as diamonds and gold, to sharks teeth and fossils. And when you get tired of digging, there are almost 200 other places to explore including museum, caves, historic points of interest and national monuments. We are certain that we have only scratched the surface. At the end of this book there is a page for you to help us make additions to a future edition. If you know of a mineral site or mine we have left out, drop us a note and we'll add it to the next edition.

Some places listed are best suited for beginners. They have instructors, equipment rentals and help for you when you need it. They are established mines with reliable, helpful and experienced staffs. We like these places because you can go and get help if you need it. They are well suited to families. There are people around with whom you can swap stories and specimens.

Some of the places listed are wild and isolated. These locations are better suited to prospectors with experience and their own equipment. They are off the beaten path and offer opportunities to enjoy the outdoors undisturbed.

This book is intended as a resource for your next trip. Pack up your car or mobile home and plan an adventure. Many places allow you to camp right at the site. Mining and collecting sites are organized to give you a good idea of what each establishment has to offer. Choose the ones that best fit your level of hunting. We have included brief information and what equipment you need to bring. PLEASE CALL AHEAD TO BE SURE THE ESTABLISHMENT WILL BE OPEN DURING THE TIME YOU PLAN TO VISIT.

We hope this guidebook will become a valuable resource to you, and that you and your family will have wonderful adventures together. We are planning a trip across North America in the near future. We will be using our own book as a reference. Maybe we will see you at some of the collecting and prospecting areas. Best of luck in your search.

Anyone who plans to visit any of these locations for the purpose of collecting or mining is advised to call ahead to make reservations and verify that the site or mine is open. Phone numbers, websites and other contact information changes constantly, but every effort for this edition was made to give the most current information possible. Some locations open and close based on the weather or snow cover. It is always a good idea to check before you make the trip.

TIPS AND RULES TO FOLLOW:

1. Let a friend know exactly where you plan to go and when you plan to return.

2. Be sure to check in when you get back. This is always a good idea and an especially important practice when searching in a remote area of the country.

3. If your search takes you off the beaten path or into an area of the country with which you are unfamiliar, do your research.

4. Don't travel alone.

5. If you need a permit or to gain permission to mine on private land, get it before you venture out. There is nothing as frustrating as getting to your destination without proper permits.

6. Be sure to bring a stocked first aid kit with you to care for cuts and bruises. In some areas of the country a snake bite kit is recommended.

7. As with any form of exercise, it is wise to check with your physician before undertaking a strenuous outing.

8. Be familiar with local plants and animals, weather conditions and climate. All this information will help you to be better prepared and insure a safe, productive expedition.

9. Check that you have packed the proper tools and equipment for your adventure.

   a. Include eye protection for mining that requires the use of hammers or chisels.

   b. Pack work gloves to keep your hands from blistering.

   c. The clothes you choose to pack will depend on the season, climate and terrain in which you will mine.

10. There are some items that are needed for most trips.

    a. Bring a hat, sturdy boots and comfortable clothes.

    b. Long pants will protect your legs from brush and other hazards.

    c. Take bug spray and sun screen with you.

    d. Bring clothing for changes in temperature and weather. Dress in layers so you can remove your windbreaker, sweater, sweatshirt or vest as the day or you heat up. Older clothing is preferable since it may get stained.

    e. A rain poncho should be in your standard equipment, as well as a clean change of clothing, socks and a towel. Dry, clean clothes cheer the spirit, soothe the soul and make the ride home more bearable.

    f. Some locations require only a swimsuit and towel, while other locations may require a wetsuit and scuba tank, so check with your destination or details.

    g. Many sites provide food and drink. Others do not. It is wise to always carry drinking water with you, especially if you are traveling in desert areas. Pack plenty of food. Physical labor builds a big appetite!

MINER'S TEN COMMANDMENTS

The following commandments were taken from the website of the Museum of the City of San Francisco who credits its origin with a man named James M. Hutchings. He wrote them in 1853 and first published them in a newspaper called the *Placerville Herald*. They were so popular, that he self-published them on letter sheets and sold them individually.[1]

A man spake these words, and said: I am a miner, wandering "from away down east," to sojourn in a strange land. And behold I've seen the elephant, yea, verily, I saw him, and bear witness, that from the key of his trunk to the end of his tail, his whole body hath passed before me; and I followed him until his huge feet stood before a clapboard shanty; then with his trunk extended he pointed to a candle-card tacked upon a shingle, as though he would say Read, and I read the:

MINER'S TEN COMMANDMENTS

I. Thou shalt have no other claim than one.

II. Thou shalt not make unto thyself any false claim, nor any likeness to a mean man, by jumping one: for I, a miner, am a just one, and will visit the miners around about, and they will judge thee; and when they shall decide, thou shalt take thy pick, thy pan, thy shovel and thy blankets with all thou hast and shall depart seeking other good diggings, but thou shalt find none. Then when thou hast paid out all thy dust, worn out thy boots and garments so that there is nothing good about them but the pockets, and thy patience is like unto thy garments, then in sorrow shall thou return to find thy claim worked out, and yet thou hath no pile to hide in the ground, or in the old boot beneath thy bunk, or in buckskin or in bottle beneath thy cabin, and at last thou shalt hire thy body out to make thy board and save thy bacon.

III. Thou shalt not go prospecting before thy claim gives out. Neither shalt thou take thy money, nor thy gold dust, nor thy good name, to the gaming table in vain; for monte, twenty-one, roulette, faro, lansquenet and poker, will prove to thee that the more thou puttest down the less thou shalt take up; and when thou thinkest of thy wife and children, thou shalt not hold thyself guiltless—but insane.

IV. Thou shalt not remember what thy friends do at home on the Sabbath day, lest the remembrance may not compare favorably with what thou doest here. Six days thou mayst dig or pick; but the other day is Sunday; yet thou washest all thy dirty shirts, darnest all thy stockings, tap thy boots, mend thy clothing, chop the whole week's firewood, make up and bake thy bread, and boil thy pork and beans, that thou wait not when thou returnest from thy long-tom weary. For in six days' labor only thou canst do it in six months; and thou, and thy morals and thy conscience, be none the better for it; but reproach thee, shouldst thou ever return with thy worn-out body to thy mother's fireside.

V. Though shalt not think more of all thy gold, and how thou canst make it fastest, than how thou will enjoy it after thou hast ridden rough-shod over thy good old parents' precepts and examples, that thou mayest have nothing to reproach thee, when left ALONE in the land where thy father's blessing and thy mother's love hath sent thee.

VI. Thou shalt not kill; neither thy body by working in the rain, even though thou shalt make enough to buy physic and attendance with; nor thy neighbor's body in a duel, or in anger, for by "keeping cool,"thou canst save his life and thy conscience. Neither shalt thou destroy thyself by getting "tight," nor "stewed," nor "high," nor "corned," nor "half- seas over," nor "three sheets in the wind," by drinking smoothing down—"brandy slings," "gin cocktails," "whiskey punches," "rum toddies," nor "egg-noggs." Neither shalt thou suck "mint juleps," nor "sherry-cobblers," through a straw, nor gurgle from a bottle the "raw material," nor take "it straight" from a decanter; for, while thou art swallowing down thy purse, and the coat from off thy back thou art burning the coat from off thy stomach; and if thou couldst see the houses and lands, and gold dust, and home comforts already lying there—"a huge pile"—thou shouldst feel a choking in thy throat; and when to that thou addest thy crooked walkings thou wilt feel disgusted with thyself, and inquire "Is thy servant a dog that he doeth these things!" Verily, thou shalt say, "Farewell, old bottle, I will kiss thy gurgling lips no more; slings, cocktails, punches, smashes, cobblers, nogs, toddies, sangarees and juleps, forever farewell. Thy remembrance shames one; henceforth, I cut thy acquaintance, and headaches, tremblings, heart-burnings, blue devils, and all the unholy catalogue of evils that follow in thy train. My wife's smiles and my children's merry-hearted laugh, shall charm and reward me for having the manly firmness and courage to say NO. I wish thee an eternal farewell."

VII. Thou shalt not grow discouraged, nor think of going home before thou hast made thy "pile," because thou hast not "struck a lead," nor found a "rich crevice," nor sunk a hole upon a "pocket," lest in going home thou shalt leave four dollars a day, and going to work, ashamed, at fifty cents, and serve thee right; for thou knowest by staying here, thou mightst strike a lead and fifty dollars a day, and keep thy manly self respect, and then go home with enough to make thyself and others happy.

VIII. Thou shalt not steal a pick, or a shovel, or a pan from thy fellow-miner; nor take away his tools without his leave; nor borrow those he cannot spare; nor return them broken, nor trouble him to fetch them back again, nor talk with him while his water rent is running on, nor remove his stake to enlarge thy claim, nor undermine his bank in following a lead, nor pan out gold from his "riffle box," nor wash the "tailings" from his sluice's mouth. Neither shalt thou pick out specimens from the company's pan to put them in thy mouth or pocket; nor cheat thy partner of his share; nor steal from thy cabin-mate his gold dust, to add to thine, for he will be sure to discover what thou hast done, and will straightaway call his fellow miners together, and if the law hinder them not, will hang thee, or give thy fifty lashes, or shave thy head and brand thee, like a horse thief, with "R" upon thy cheek, to be known and read of all men, Californians in particular.

IX. Thou shalt not tell any false tales about "good diggings in the mountains," to thy neighbor that thou mayest benefit a friend who had mules, and provisions, and tools and blankets he cannot sell,—lest in deceiving thy neighbor, when he returneth through the snow, with naught save his rifle, he present thee with the contents thereof, and like a dog, thou shalt fall down and die.

X. Thou shalt not commit unsuitable matrimony, nor covet "single blessedness;" nor forget absent maidens; nor neglect thy "first love;"—but thou shalt consider how faithfully and patiently she awaiteth thy return; yea and covereth each epistle that thou sendest with kisses of kindly welcome–until she hath thyself. Neither shalt thou cove thy neighbor's wife, nor trifle with the affections of his daughter; yet, if thy heart be free, and thou dost love and covet each other, thou shalt "pop the question" like a man.

A new Commandment give I unto thee—if thou has a wife and little ones, that thou lovest dearer than life,–that thou keep them continually before thee, to cheer and urge thee onward, until thou canst say, "I have enough–God bless them–I will return." Then from thy much-loved home, with open arms shall thy come forth to welcome thee, with weeping tears of unutterable joy that thou art come; then in the fullness of thy heart's gratitude, thou shalt kneel together before thy Heavenly Father, to thank him for thy safe return. AMEN—So mote it be.

FORTY-NINER[1]

Footnote
[1] http://www.sfmuseum.org/hist7/tencom.html, Gladys Hansen, Museum of the City of San Francisco, 2001.

Now you need to pack your tools. The following list consists of the standard tools you will need for almost all of your adventures. This list is meant to help you get organized.

STANDARD TOOLS:
- 5-gallon Bucket
- Chisels
- Closable Plastic Container (Film Canister)
- Collecting Bag
- Crowbar/Pry bar
- Eye Protection (Goggles)
- First-aid Kit
- Gads
- Screwdriver
- Rock Hammer
- Pencil and Notepad
- Pick
- Proper Clothing
- Rags
- Shovel
- Sledgehammer
- Small Trowel and Hand Garden Tools
- Snacks
- Water
- Work Gloves
- Ziplock Bags

Some mines require specialized equipment. Whenever possible we have tried to include that information with the listing for the individual mine.

ADDITIONAL EQUIPMENT FOR SAPPHIRE / DIAMOND / OPAL MINING:
- Plastic Spray Bottle
- 1/8-, 1/4- and 1/2-inch Sifting Screens
- Tweezers
- Washtub
- Whisk Broom or Brush

GOLD MINING TOOLS:
- Collecting Bottles
- Gold Pan
- Snuffer Bottle

ADDITIONAL EQUIPMENT FOR EXPERIENCED GOLD MINERS:
- Dredge
- Highbanker
- Metal Detector
- Sluice Box

If you plan to do any metal detecting, you need to pack your detector, a pouch to hold your finds and some kind of scoop or shovel. Many mines can provide all the tools that you will need. Some rent equipment, while still others provide very little. We suggest that you check with the individual mine if you are uncertain of what to bring along.

To make the most out of your adventure once you get there you need to observe the following words to the wise to give you a leg up on the greenhorns.

1. When swinging rock hammers, sledgehammers or a pick, make sure everyone is clear of your work area and that you have adequate footing.

2. Never enter caves or old mining tunnels. Snakes love caves for their cool shade. Caves are also frequented by other animals including mountain lions and bears. Old shafts are extremely dangerous and unstable. They are not worth the risk. Never tunnel into banks with overhangs or on cliffs.

3. If you are searching for gold for the first time, try to get a panning lesson. Many mines teach gold panning. You can teach yourself of course, but, in our opinion, nothing compares to watching an expert and gaining some real experience.

4. Stay out of fast-moving water. Do not damage or prospect in the riverbank or disturb plant life while mining. Be sure you are not in a flash flood area. Heavy rains upstream can flood downstream in minutes and under a clear blue sky.

5. Check out the plant life on the banks to avoid poison ivy or other harmful plants.

6. Contain all camp fires in a fire pit.

7. If you are working a dredge, clear all large rocks away from the hole you are working to prevent cave-ins or rock slides. Be careful when digging into a bank that the hole does not collapse. Never work in a hole deeper than you are tall if you are alone.

8. Remember to fill in the holes you dig. Holes are dangerous and unsightly.

9. Watch over children and inexperienced miners. Instruct children on how to use tools. Point out hazards and take frequent breaks.

The scales used for precious metals and gemstones are different than those to weigh zucchini. Gold miners speak on pennyweight and jewelers talk in terms of carats and grams.

Gems are weighed in carats, not to be confused with karats used to distinguish the purity of gold. The word carat comes from ancient India where carob seed were used as small consistent weights.

The word pennyweight comes from the Gold Rush days when a miner would compare the weight of his gold dust in comparison to the weight of a penny. The size of a penny in 1849 was somewhat larger than the flimsy zinc versions minted now.

The size of gemstones is measured in millimeters.

Here are some conversions to help you see what is what.

PENNYWEIGHT CONVERSIONS:
- 1 Pennyweight (dwt) = 7.776 Carats
- 1 Pennyweight (dwt) = 1.55 Grams
- 1 Pennyweight (dwt) = 24 Grains
- 1 Pennyweight (dwt) = 0.05 Troy Ounces

CARAT CONVERSIONS:
- 1 Carat = 0.20 Grams
- 1 Carat = 0.1287 Pennyweight
- 1 Carat = 0.0064 Troy Ounce
- 1 Gram = 5 Carats
- 1 Pennyweight = 7.776 Carats
- 1 Troy Ounce = 155.5 Carats

TROY WEIGHT
- 24 Grains = 1 Pennyweight
- 20 Pennyweight = 1 Troy Ounce
- 12 Troy Ounces = 1 Troy Pound

- 1 Millimeter = 0.03937 Inches
- 1 Inch = 25.4 Millimeters

MOHS TABLE OF HARDNESS
Friedrich Mohs devised the scale of mineral hardness in 1822. He picked ten minerals and rated them in terms of their strength with ten being the hardest. A Mohs criterion for a mineral to be listed with a higher number was that it could scratch the specimen in the category below it. Mohs original representative minerals are underlined in the table below.

MOHS SCALE

1	<u>Talc</u>
2	<u>Gypsum</u>, Pearl
3	<u>Calcite</u>, Malachite
4	<u>Fluorite</u>, Opal
5	<u>Apatite</u>, Glass, Lapis
6	<u>Orthoclase</u>, Jade, Hematite, Onyx, Peridot, Zircon
7	<u>Quartz</u>, Garnet, Emerald, Amethyst, Aquamarine, Citrine, Tourmaline
8	<u>Topaz</u>, Chrysoberyl, Cubic Zirconia
9	<u>Corundum</u>, Ruby, Sapphire
10	<u>Diamond</u>

The Reed Gold Mine reported the first gold find in the United States. As the story goes, in 1799, Reed's son, Conrad, found a large yellow rock in Little Meadow Creek while playing hooky from church. A silversmith in Concord was unable to identity the rock, but measured its weight at seventeen pounds. The Reeds used it as a door stop until a jeweler bought it for $3.50 in 1802, less than one-tenth of its value. Later Reed discovered his mistake and "convinced" the jeweler to compensate him. Mining later began by Reed and others and a twenty-eight-pound nugget was found. A total of $100,000 worth of gold was unearthed by 1824. North Carolina was the first gold producing state in the United States

Gold was discovered in 1806 near Spotsylvania County, Virginia. Several mines were in operation by 1825. Mining ceased in 1849 when the Virginia miners left the state to head for the gold rush in California. In 1850, the production of gold in Virginia was reduced by half because of a labor shortage. Gold was last mined commercially in Virginia in 1947 as a by-product of lead mining.

Gold was also discovered in South Carolina in the early 1800s. Lancaster County was shipping gold to the U.S. Mint in 1829. Mining, interrupted by several wars, continued throughout the 1800s.

Georgia had a gold rush in 1828 which lasted to 1850 with over four thousand miners working the hills. Gold was first discovered in Georgia in 1818 by Mr. Benjamin Hicks. He made the discovery by kicking over a rock to expose a lump of gold the color and size of an egg's yolk. Miners deserted this area as their counterparts in Virginia had, when gold was discovered in California. But, they came back when California gold ran out in 1855. Mining was again interrupted, this time by the Civil War, to be resumed again after the end of the war. Commercial mining of gold continued until World War II.

New Mexico gave the cry of gold in the Ortiz Mountains, south of Santa Fe, in 1828. This was not the first gold rush in this area. Spanish conquistadors discovered and mined gold here after searching throughout South America and Mexico. Prior to that Native Americans mined gold here for unknown periods of time.

Alabama had its own gold rush beginning in the 1830s. Gold was discovered in Tennessee in 1831 on the Ococee Land District, which belonged to the Cherokees. The land was taken along with the gold. The most famous area here was Cokercreek, which is still producing gold today.

In 1838, gold was found in Ohio where it was deposited fourteen thousand years earlier during the glacier's retreat. The origin of this gold was probably somewhere in Northern Canada.

The largest gold rush in North America occurred in 1849 in California. Gold had been discovered a year earlier at Sutter's Sawmill by a worker named James W. Marshall. In January of 1848, a dam was built on the American River. Water was channeled to the dam to remove the loose dirt and gravel and then turned off again.

On Monday, January 24, 1848, James Marshall walked in the channel and spotted grains of shiny metal the size of wheat. Suspecting this was gold he rushed to tell his fellow workers that he had found a gold mine. Captain John Sutter was informed of the potential find and did all within his power to prevent this story from leaking out, while buying as much surrounding land from the Native Americans as possible. The remoteness of the area slowed the speed of the discovery.

In March of 1848, gold was mined and the discovery reported in the San Francisco newspapers. Soon men from around the world began to migrate to Coloma, California to establish the first mining camp. John Sutter was there to sell supplies, food, tools and claims to the would-be miners.

In the fall of 1849, the number of miners in California had forever changed the population distribution in the United States. Men worked in groups, mining an area until the gold disappeared. The miners then moved up the creeks and rivers to search for other gold deposits. Camp towns sprung up overnight, to disappear just as quickly. Many of these towns are well preserved and still exist along Highway 49. Their names create vivid images; Fiddletown, Rough and Ready, Chinese Camp, Sutters Creek, Placerville and Gold Run—to name a few. In 1853, at least $65 million worth of gold was taken from California mines. Mining in California continued until the surface gold played out in 1855.

During the same time period, gold was discovered in Alaska at the Kenai River in 1848 by a Russian mining engineer. Alaska did not become a U.S. territory until March 30, 1867. Alaska had a series of strikes or rushes in Anchorage, Nome and Fairbanks. Gold continues to be discovered today in Alaska.

Nevada was found to have gold in 1849. During the gold mining process, much blue clay was uncovered and discarded. This clay was later found to be rich in silver. Silver became the true treasure of Nevada.

The western states were next searched for gold and silver. It was found in 1852 in Idaho, Montana and Oregon, in Washington in 1853 and Utah in 1858. That same year gold miners in Arizona were earning $4 to $150 a day.

In 1859, the slogan "Pike's Peak or Bust" became the cry of gold miners heading for Colorado. Gold production in Colorado reached forty million troy ounces.

Montana, which had shown some gold in 1852, became the next rush site in 1863, when large quantities of gold were found near Virginia City.

The Black Hills of South Dakota, in and around Rapid City, experienced their own influx of miners for gold between 1876 and 1878.

Gold is still mined commercially in the U.S. and small scale miners are still out there, even as you read this, with picks, shovels, dredges and sluice boxes looking for that precious metal.

 Interstate Highway

 U.S. Highway

 State Highway

 City

 State Capital

 River

 National Park, Monument, Forest, Wilderness, Wildlife Refuge

 Site Number Marker

State Diggings Site Locator Map

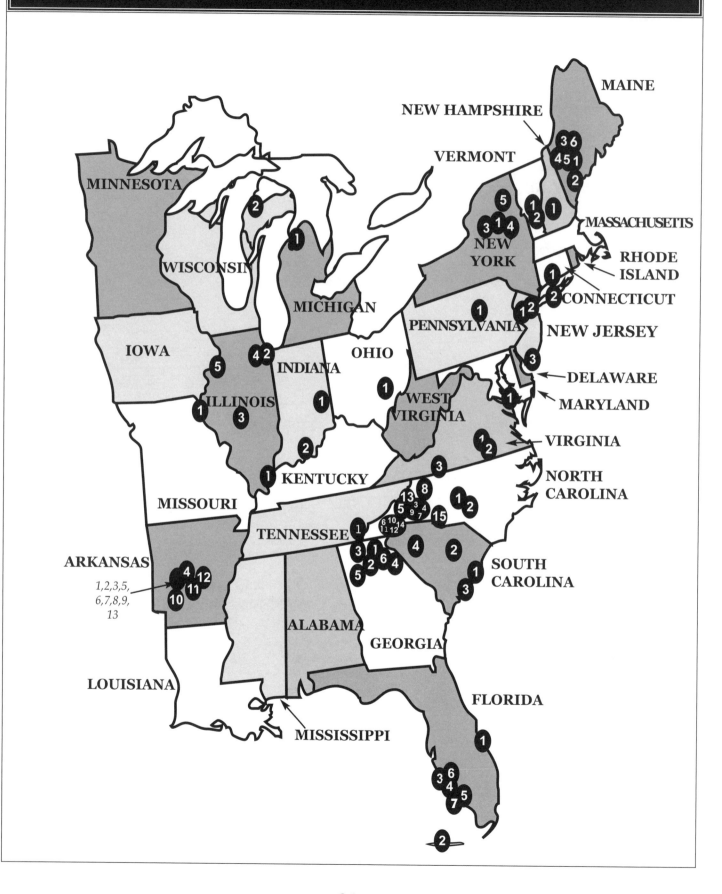

MAINE

NEW HAMPSHIRE

VERMONT

MINNESOTA

MASSACHUSETTS

RHODE ISLAND

CONNECTICUT

NEW YORK

WISCONSIN

MICHIGAN

PENNSYLVANIA

NEW JERSEY

IOWA

OHIO

DELAWARE

MARYLAND

INDIANA

ILLINOIS

WEST VIRGINIA

VIRGINIA

NORTH CAROLINA

KENTUCKY

MISSOURI

TENNESSEE

ARKANSAS

SOUTH CAROLINA

1,2,3,5, 6,7,8,9, 13

ALABAMA

GEORGIA

LOUISIANA

FLORIDA

MISSISSIPPI

26

ARKANSAS
- (1) Crystal Heaven
- (2) Twin Creek Mine & Rock Shop
- (3) The Crystal Seen Trading Co.
- (4) Gee & Dee Crystal
- (5) The Starfire Mine
- (6) Wegner Crystal Mines & Ranch
- (7) Arrowhead Mining
- (8) Fiddler's Ridge Rock Shop & Crystal Mine
- (9) Sonny Stanley's
- (10) Crater of Diamonds State Park
- (11) Coleman's Quartz Mine
- (12) Willis Crystal Mine
- (13) Sweet Surrender Crystal Mines

CONNECTICUT
- (1) Green's Garnet Farm

FLORIDA
- (1) Vero Beaches
- (2) Atocha/Margarita Expeditions
- (3) Caspersen Beach
- (4) Giant Stride
- (5) Wild Tours/Fossil Expeditions, Megalodon Expeditions, and Everglade Eco-tours
- (1) Venice Public Beaches
- (2) Sanibel and Captiva Islands

GEORGIA
- (1) Consolidated Gold Mine
- (2) Crisson Gold Mine
- (3) Hackney Farm
- (4) Gold 'N Gem Grubbin'
- (5) Graves Mountain
- (6) Hidden Valley Campground

INDIANA
- (1) Yogi Bear's Jellystone Park/Knightstown
- (2) Interstate 64 Roadcut

MAINE
- (1) Perham's of West Paris
- (2) Poland Mining Camp
- (3) Bemis Stream Prospect
- (4) Bumpus Quarry
- (5) Songo Pond Mine
- (5) Swift River Gold Panning Area

MARYLAND
- (1) Calvert Cliffs State Park

MICHIGAN
- (1) Petoskey State Park
- (2) Delaware Copper Mine

MISSOURI
- (1) Sheffler's Rock Shop

NEW HAMPSHIRE
- (1) Ruggles Mine

NEW JERSEY
- (1) Sterling Hill Mining Museum
- (2) Franklin Mineral Museum & Mine Replica
- (3) Cape May Public Beach

NEW YORK
- (1) Treasure Mountain Diamond Mine
- (2) Ace of Diamonds Mine
- (3) Herkimer Diamond Mine
- (4) Crystal Grove Diamond Mine and Campground
- (5) The Barton Mine

NORTH CAROLINA
- (1) Cotton Patch Gold Mine
- (2) Reed Gold Mine
- (3) The Lucky Strike Gold and Gem Mine
- (4) Heather Grove Mine
- (5) Old Pressley Sapphire Mine
- (6) Mason Ruby and Sapphire Mine
- (7) Thermal City Gold Mine
- (8) Emerald Hollow Mine
- (9) Vein Mountain Gold and Gems
- (10) Rose Creek Mine
- (11) Sheffield Mine
- (12) Old Cardinal Gem Mine
- (13) Emerald Village
- (14) Mason Mountain Rhodolite and Ruby Mine
- (15) Golden Valley Campground

OHIO
- (1) Hidden Springs Ranch

PENNSYLVANIA
- (1) Crystal Point Diamond Mine

SOUTH CAROLINA
- (1) Cooper River Divers
- (2) Deep South Rivers
- (3) Edisto Beach
- (4) William's Property

TENNESSEE
- (1) Burra Burra Mine

VERMONT
- (1) Richard's Gold Mine–Gold Panning
- (2) Camp Plymouth State Park–Gold Panning

VIRGINIA
- (1) Morfield Gem Mine
- (2) Deck Boyles Farm and Mine
- (3) Fairy Stones State Park

Pearls, Iron and Coal

In November of 1833 the earth orbited through the path of the Tempel – Tuttle Comet as it does every year, causing a meteor shower. These showers are usually lovely, but not generally noticed by the public. Every 33 years, however, the earth crosses nearest the comet tail. This was the case in 1833 and the spectacular celestial event became known as "the night stars fell on Alabama." Instead of the usual 10 to 20 shooting stars per hour, the night was ablaze with an estimated 24,000. Many Alabamans woke from their beds to the unfamiliar light. Some thought they witnessed the end of the world.

This same year a spectacular fossil was unearthed on a plantation in southwest Alabama. Early reports were that the bones came from a giant reptile dubbed Basilosaurus (king of lizards). The fossil's total length is estimated at 55 to 70 feet. Further study revealed that the creature was actually a mammal. Teeth and jaw indicated this was a primitive meat-eating whale.

A man named Albert Koch came to Alabama in 1845 in search of a sea serpent of his own and found one, which he shipped to New York via Key West, where the cargo wrecked on the reefs. Salvagers thought the discovery important enough to rescue, and sent the fossils to New York free of charge. Instead of reconstructing the bones, as they would be found in a whale, Koch strung them like a sea monster or dragon. His monster drew great crowds and much criticism from paleontologists.

Back home in Alabama, gold was discovered in Tallapoosa County. No one knows the exact person or place where gold was first found. Farmers seemed to know of the gold and Native Peoples were undoubtedly aware of the metal's presence. By the mid-1830s gold mining operations appeared. Miners sluiced and high banked the gravel along gold-bearing streams while searching for quartz veins laced with gold. One man, Dr. Ulrich, unearthed gold while digging a wine cellar. Digging continued until 1849 when the call of "Gold" came from California. Miners left their stakes in droves, heading west. By the 1850s coal mining was more profitable than gold.

Red hematite (iron ore) was discovered in abundance in the area now known as Birmingham. By the 1860s the Alabama iron and steel industry was born. Alabama furnished the confederacy with iron for cannons, rifles, and naval armaments.

The city of Birmingham is proud of its roots as forgers of iron. To commemorate their past, the city commissioned the building of the largest cast-metal statue in the world. The statue was titled "The Iron Man" and is modeled after the Roman smithy god Vulcan, the God of fire, jewels and weapons. The father of blacksmithing was believed to be ugly and hard working. His wife, Venus, the Goddess of Love and Beauty, more than compensates his lack of physical beauty. The good citizens funded the creation of the 56-foot tall statue of Vulcan for the St. Louis Worlds Fair. The 120,000-pound statue holds a hammer in one hand and a spear point in the other. He now stands in his own park on top of Red Mountain.

Vulcan would be proud of another of Alabama's natural resources. His wife Venus arose from the sea and is often depicted emerging from the shell of an oyster. Oysters are big business in Alabama. Fresh water pearls are harvested in Alabama. These tiny miracles are valued for their unusual shape and high luster. The mussels were gathered for the meat and the shells used to make buttons until the 1950s when manufacturers switched to plastic. The industry fell on hard times until the 1970s when the Japanese developed a technique to make cultured pearls using the shell of freshwater oysters as the seed for the ocean oyster's pearl.

Today hobbyists still search for gold in the gold-bearing areas of Alabama. This state also has many show caves with geologic beauty and historical significance. The museums in Alabama highlight some of the states finest fossils including a Mosasaurus on display at the Alabama Museum of Natural History.

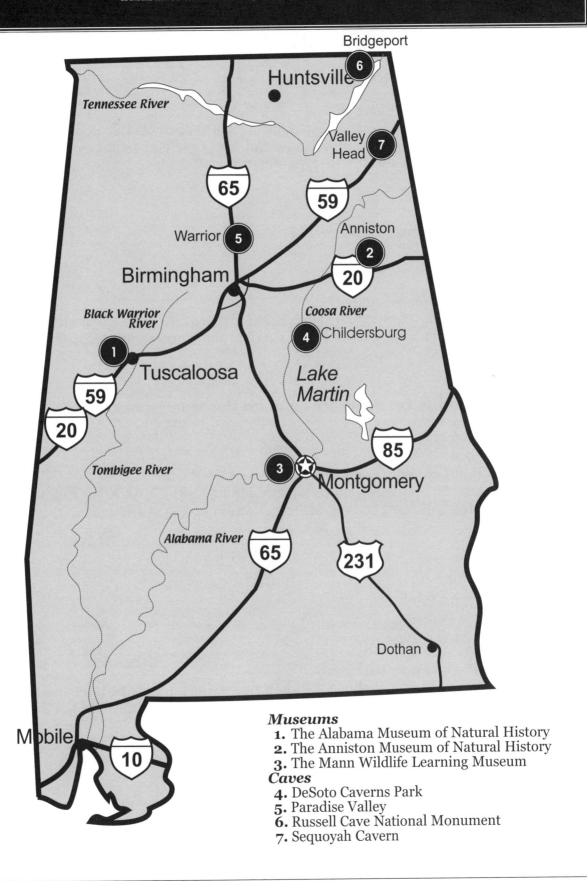

Museums
1. The Alabama Museum of Natural History
2. The Anniston Museum of Natural History
3. The Mann Wildlife Learning Museum

Caves
4. DeSoto Caverns Park
5. Paradise Valley
6. Russell Cave National Monument
7. Sequoyah Cavern

MUSEUMS

Site 1.
The Alabama Museum of Natural History
Box 870340
Tuscaloosa, AL 35487-0340
(205) 348-7550
http://museums.ua.edu/history/
This museum is located on the campus of the University of Alabama and includes displays entitled the Age of Dinosaurs, Coal Age and Ice Age. The museum also has an extensive mineralogical collection and the Hodges Meteorite, which is the only meteorite known to have struck a human.

Site 2.
The Anniston Museum of Natural History
P.O. Box 1587
Anniston, AL 36202-1587
(256) 237-6766
http://www.annistonmuseum.org/
A permanent exhibition called Dynamic Earth explores the forces that shaped our planet and includes fossils and a geology collection. There is also a recreation of an Alabama cave right in the building.

Site 3.
The Mann Wildlife Learning Museum (at the Montgomery Zoo)
P.O. Box 3242
Montgomery, AL 36110
(334) 240-4900
http://www.mannmuseum.com/
The Mann Museum is largely a collection of dioramas with North American Animals, including bear, moose, elk, and wolf. There is also a collection of fossils and artifacts.

CAVES

Site 4.
DeSoto Caverns Park
DeSoto Caverns Parkway
Childersburg, AL 35044
(800) 933-2283
http://www.desotocavernspark.com/
This cavern was used as a shelter for Indian traders as far back as 1723 and is one of the first officially recorded caves in the U.S. in 1796. The Confederate Army used the location and so did moonshiners during Prohibition.

Site 5.
Rickwood Caverns
370 Rickwood Park Road
Warrior, AL 35180-9803
(205) 647-9692, (205) 647-9692, or (800) ALAPARK
http://www.dcnr.state.al.us/parks/rickwood_more_info.htm
This cave presents a mile of living geology, which continues to form over its 260-million-year history. The "miracle mile" includes passages and lighted chambers with thousands of limestone formations.

Site 6.
Russell Cave National Monument
3729 County Road 98
Bridgeport, AL 35740
(256) 495-2672
http://www.nps.gov/ruca/
This National Monument contains archeological treasures of prehistoric people living in the Southeast nearly 10,000 years ago. Evidence of indigenous tribes includes clothing, food stuffs and weaponry.

Site 7.
Sequoyah Cavern
1438 County Road 731
Valley Head, AL 35989
(800) 843-5098 or (256) 635-0024
http://www.alabamatravel.org/north/scc.html
This tour includes a description of the cave's historical figures; Chief Sequoyah, for whom the cave is named and Sam Houston. Visitors will see a unique lake as still and reflective as glass. Aboveground guests can enjoy a campsite with unusual pets including buffalo, deer, goats and lamb.

The only place in the world the public can dig for and keep all the diamonds they find.

Arkansas is a rock hunter's paradise, and if you love fossils this state will not disappoint. Beginning with the most abundant mineral – Quartz, the town of Mount Ida must be introduced. Rockhounds have journeyed here for years to visit the plethora of crystal mines clustered, like the quartz itself, in this small town. Visitors arriving in October may take part in the Annual World's Championship Quartz Crystal Dig. At other times of the year diggers will find lovely clear and cloudy quartz crystal points and clusters of all description.

Have you heard of the Crater of Diamonds State Park? This is the only place in the world that allows the public to search a kimberlite pipe for diamonds – real diamonds. Over 60,000 visitors a year search for this most precious carbon. Reports indicate that 600 or so diamonds are found per annum. Perhaps you have a one in ten shot at a diamond and a nine in ten shot of going home with nothing but dirty blue jeans. Most diamonds are recovered by hours of backbreaking digging and sifting. Some are found by just strolling along with an eye to the ground. That is how a guest found the Amarillo Starlight, the largest diamond ever recovered by a park visitor. The clear white diamond, weighing over 16 carats, was spotted after a rainstorm. The gem was said to be the size of a wild pecan.

The Crater itself has been known to yield diamonds for over 90 years. The original owner, a farmer named John Huddleston, found a clear stone and a translucent yellow stone in 1907 while clearing a field. When the stone put a groove in his grinding wheel he went to a local banker with his discovery. Tiffany & Company verified the stones as diamonds. Huddleston sold the land for $36,000 cash, which he figured would last him and his family a lifetime. It did not.

Commercial mining began in 1919, but the mine was wrought with sabotage and employees who stole the diamonds nearly as fast as they were discovered. The mine failed and the land was sold. A new owner had the soil tested, finding many fewer diamonds in the sample than expected. Eventually the area became a tourist attraction and later a state park. The largest gem ever found here is the Uncle Sam, unearthed in 1924 and weighing 40.23 carats. The second largest diamond discovered is the Star of Murfreesboro at 34.25 carats.

But Arkansas mining history includes more than diamonds. Before their discovery, Arkansas was known for coal, zinc and lead mines. As far back as 1818 lead for bullets was mined in northern Arkansas. During the Civil War this area had three lead smelting furnaces to meet the demand for ammunition. Zinc, mined from 1857, was used with other metals to make stronger alloys and to coat steel and iron to prevent corrosion. One of the stranger uses of zinc occurred before the discovery of antibiotics when zinc sulfate was used, with marginal success, to treat venereal disease.

All that zinc mining has left behind tailings piles with some beautiful dolomite samples. Iron and aluminum mining unearth lovely wavellite specimens. Arkansas rockhounds have connections to get into working mines and know where to find tailings piles from abandoned mines.

Beneath the soil, Arkansas' base rock is largely sedimentary. Over half of this state was once submerged beneath an inland sea. Limestone quarries preserve marine fossils including sharks' teeth, the remains of cephalopods like octopus and squid, and free-floating animals like nautilus. Fish, ray, and primitive whale have also left remains here. In other areas, where land was never submerged, diggers find plant remains, such as ferns. Dinosaurs are occasionally found, such as the state official dinosaur, the Arkansaurus fridayi. This aggressive carnivore may have hunted with others of its species, as did velociraptors. This dinosaur has a mystery. Scientists do not know if the creature has teeth like a raptor or a horn-like bill like an eagle because no teeth have been recovered to solve the puzzle.

So what are you waiting for? Shouldn't you be planning a trip to Arkansas?

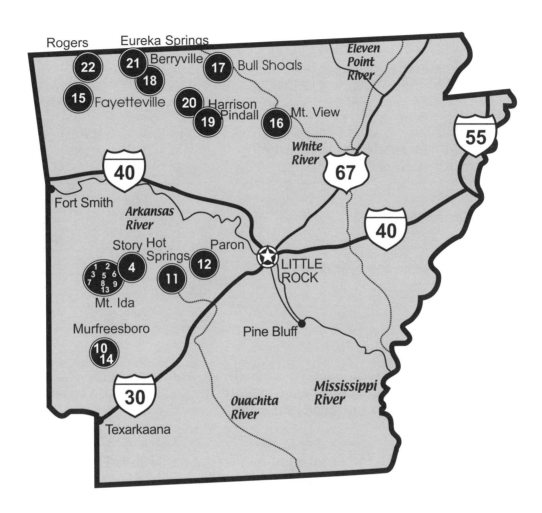

Diggings
1. Crystal Heaven
2. Twin Creek Mine
3. The Crystal Seen Trading Co.
4. Gee & Dee Crystal
5. The Starfire Mine
6. Wegner Crystal Mines and Ranch
7. Arrowhead Crystal Mine
8. Fiddler's Ridge Rock Shop and Crystal Mine
9. Stanley Rock Shop
10. Crater of Diamonds State Park
11. Coleman's Quartz Mine
12. Willis Crystal Mine
13. Sweet Surrender Crystal Mines

Museums
14. Crater of Diamonds State Park
15. University Museum

Caves
16. Blanchard Springs Cavern
17. Bull Shoal Caverns
18. Cosmic Caverns
19. Hurricane River Cave
20. Mystic Caverns
21. Onyx Cave Park
22. War Eagle Caverns

ADDRESS:
Crystal Heaven
39 Sunrise Hills Drive
Mount Ida, AR 71957
crystall@ipa.net
www.crystalheaven.com
(870) 867-5116

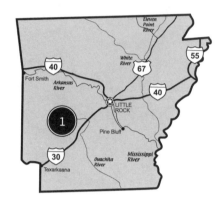

DIRECTIONS:
Crystal Heaven Mine is located 4 miles east of
Mount Ida on Sunrise Hills Drive.

SEASON:
May 1 to late November

HOURS:
8:30 a.m. to 3:00 p.m. by appointment only.

COST:
$20 per person per day.

WHAT TO BRING:
Hand tools such as picks, long handled screwdriver, rock hammers. Bring a bucket or two, gloves and eye protection. A hat and sunscreen are advised along with newspaper to wrap your best crystal finds. Bring your own food and drink.

INFORMATION:
Guests need to supply all their own equipment, food and water. There are no facilities at this site. Material is exposed regularly and there are guides available to help first timers.
Take your time while digging so you do not damage the fragile points. The owners say walk-ins are welcome, but it might be best to call ahead.

ADDRESS:
Crystal Rockshop
Highway 270 East, Box 547
Mt. Ida, AR 71957
info@crystal-rockshop.ch
http://www.crystal-rockshop.com/portrait_usa.html *(in German)*
(870) 867-2975

DIRECTIONS:
Go to the rockshop in Mt. Ida located on Highway 270 East. Pay your fees and get directions to the mine.

SEASON:
Open year round

HOURS:
9:00 a.m. to 5:00 p.m.
We recommend calling ahead.

COST:
Adult fee $20 per person per day
Child fee $10 per day

WHAT TO BRING:
Hand tools such as picks, long handled screwdriver, rock hammers. Bring a bucket or two, gloves and eye protection. A hat and sunscreen are advised along with newspaper to wrap your best crystal finds. Bring your own food and drink.

INFORMATION:
Stop by the owners rockshop to see specimens from around the world. This is a working mine that allows visitors to dig in older pits. It is advisable to stay clear of flagged areas and machinery.

A picnic table and restroom are available at the site. Take your time while digging so you do not damage the fragile points.

ADDRESS:
The Crystal Seen Trading Co.
2568 Hwy 270 East
Mount Ida, AR 71957
(870) 867-4072
info@crystalseen.com
http://crystalseen.tripod.com

DIRECTIONS:
Located 4 1/2 miles east of Mt. Ida in Hurricane
Grove. Located on north side of highway.

SEASON:
Open year round

HOURS:
Larger groups please call ahead or e-mail to make
reservations.

COST:
Adults - $20 per person per day for collecting at the mine.
Adults - $8 per hour or $20 per day for searching through tailings piles.
Children ages 7 - 10 $4 per hour or $10 per day searching through tailings piles.
Children 6 and under are free.

WHAT TO BRING:
The Crystal Seen Trading Co. will provide all materials for searching through the tailings piles.
A hat and sunscreen are advised. Bring your own food and drink.

INFORMATION:
    The Crystal Seen Trading Co. is the perfect place for families with children or people interested in crystal mining but don't have much time or money. The owners will work with you as you dig through the tailings to help you recognize and identify the crystals you dig up. The tailings piles come from eight different mines, so you are guaranteed to find something.

    One of the owners is an experienced miner and, and the other is a jeweler. Children of all ages can enjoy having their special find set in jewelry. Restrooms are available at the shop, and during the summer drinks and ice cream are also sold. A variety of packages is offered by the shop for the novice day visitor to the more experienced miner, call to find out more information.

ADDRESS:
Gee & Dee Crystal
4764 Highway 27, North
Story, AR 71970
(870) 867-4561

DIRECTIONS:
Take Highway 270 west to Highway 27 north.
The rockshop is located on right hand side of
highway 12 miles north of Mt. Ida.

SEASON:
Open all year round

HOURS:
Summer - 8 a.m. to sunset
Winter - 9 a.m. to 5 p.m.

COST:
$20 per person – Fee

WHAT TO BRING:
Bring standard mining tools, your own food, and plenty to drink.

INFORMATION:
Gee & Dee Crystal shop runs the Brewster Mountain Mine. People can dig all day for quartz crystals. Your $20 fee allows you to carry out as much as you can carry. You can also sign up for a "Big Dig", where for $250 you can dig inside the pocket where they are currently mining for crystals. The average haul from a "Big Dig" is 20 to 25 buckets of material.

The Brewster Mountain Mine took first place in all 10 categories in the 2003 Crystal Championship. This championship is held during the annual Quartz, Quiltz and Craftz Festival in October. Contact the Mt. Ida Chamber of Commerce at (870) 867-2723 for more information on this annual festival with activities for the entire family.

ADDRESS:
5403 Highway 270 East
Mount Ida, AR 71957
(870) 867-2431
crystals@ipa.net
http://www.www.starfirecrystals.com/

DIRECTIONS:
The Starfire Mine is located 11 miles east of Mount Ida on
U.S. Highway 270. At the time of publication, the mine is
next to the Colonial Motel.

SEASON:
Open all year round

HOURS:
By advance appointment only

COST:
$20 per person – Fee
$10 per child between ages 5 and 12 - Fee
Free – Children under 5 years of age

WHAT TO BRING:
Bring small garden tools for digging, rock hammers, a large heavy duty screwdriver or small pry
bar, a heavy duty three-pronged garden rake, protective glasses, and gloves. Some of these
items are available at the mine. Wear old clothes, sunscreen, and a hat.

INFORMATION:
The Starfire Mine is recommended for more experienced miners, due to the amount of
digging that must be done. This mine is not cleared by heavy equipment, so digging is harder
and more strenuous. The mine is known for small, bright, and clear quartz crystals. They also
have unusual stones including: tabbies, phantoms, inclusions, and faden quartz. Some good
finds have come out of the Starfire Mine. If you have experience and your own tools, it is worth
the stop. Starfire Mine also sells quartz crystals wholesale and retail to the public.

Plan to stay at least half a day, longer if you can. Crystals form from five feet to two hundred
feet down, so it may take you a while.

There is a motel associated with this mine. RV and tent camping is available nearby.

To discover more about this area you may want to write ahead of time for information from
the Mount Ida Area Chamber of Commerce at P.O. Box 6, Mount Ida, Arkansas 71957. They can
be reached by phone at (870) 867-2723. The Chamber of Commerce sponsors an annual Quartz,
Quiltz and Craftz Festival in October which you may want to attend if you are in the area. A
crystal quartz dig is part of the festivities.

ADDRESS:
P.O. Box 205
Mount Ida, AR 71957
(870) 867-2309
wegner@wegnercrystalmines.com
http://www.wegnercrystalmines.com

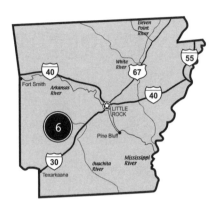

DIRECTIONS:
The Wegner Crystal Mines are located 6 miles south of Mount Ida, off State Highway 27 on Owley Road. Follow the signs on Owley Road to the mine.

SEASON:
Open all year round – Weather and holidays permitting

HOURS:
8:00 a.m. to 4:30 p.m. – Monday - Friday
Saturdays & Sundays by appointment only

COST:
$20 – Fee for Phantom Mine
$15 – Fee for Crystal Forest Mine
$8 – Fee for Old Mountain Top
$6 – Fee for Salted Mine (children and seniors)
Half-price – Fee to all the mines for children ages 12 and under

WHAT TO BRING:
Bring standard mining tools, your own food and plenty to drink.

INFORMATION:
The Wegner Crystal Mines have something for everyone. This is an extensive operation which is actually four separate mines. The Phantom Mine is the most productive and popular mine and has produced world-famous phantom crystals. The Crystal Forest Mine is located on forty acres and produces clear, gem-quality crystals. Old Mountain Top has breathtaking views and involves a strenuous climb. The mine recommends that only experienced hikers try this mine; children will have trouble making the trip. The Salted Mine is for children and seniors looking for guaranteed success. It is readily accessible and is located adjacent to the campground.

They have a tourist center which will orient you to the options available. At the center you will find a large collection of quartz crystals, minerals and fossils for sale. Snacks, sandwiches and beverages are also sold here. The mine has camping facilities with showers, restrooms, a fishing area and swimming. There is also a specimen museum and a 10,000-square-foot wholesale barn where specimens are for sale.

ADDRESS:
Arrowhead Crystal Mine
P.O. Box 1478
Mount Ida, AR 71957
(870) 326-4977

DIRECTIONS:
Take Highway 27 south approximately 3.5 miles. Turn left onto Owley Road. Continue until the pavement ends and follow sign at top of the hill.

SEASON:
Open all year round, weather permitting

HOURS:
Call ahead to make an appointment.

COST:
$25 per person

WHAT TO BRING:
Hand tools such as picks, long handled screwdriver and rock hammers. Bring a bucket or two, gloves and eye protection. A hat and sunscreen are advised along with newspaper to wrap your best crystal find. Bring your own food and drink.

INFORMATION:
This is a primitive site with no food or restrooms available. A variety of quartz crystals are found in the mines of this area including tabbies, phantoms, smoky quartz, and lovely clusters. Take your time while digging so you do not damage the fragile points.

ADDRESS:
3752 Highway 270 East
Mount Ida, AR 71957
(870) 867-2127
fecho@ipa.net
http://www.fiddlersridgecrystals.com

DIRECTIONS:
The Fiddler's Ridge Rock Shop and Mine is located 7 miles east of Mount Ida, Arkansas on U.S. Highway 270. The Crystal Mine is located approximately 5 miles from the rock shop.

SEASON:
Open all year round

HOURS:
Open daylight hours (mine)
9:00 a.m. to 4:30 p.m. (rock shop hours)

COST:
$10 per day – Adults
$5 – Children ages 10 and under

WHAT TO BRING:
Bring standard mining tools, your own food and plenty to drink.

INFORMATION:
Stop at the Fiddler's Ridge Rock Shop to get a permit to mine. We suggest you get a permit the day before, so you can set out early on your adventure. It is cool in the morning, but gets hot in the afternoon. The mine is open until dark and produces some fine crystals.

The Fiddler's Ridge Rock Shop has a large supply of mineral specimens, polished stones and unfinished stones for the rock hobbyist. Jewelry and specialty gift items are available with the native Mount Ida crystals.

Camping is available 2 miles from the shop at Lake Ouachita. Restaurants and motels are also nearby.

ADDRESS:
P.O. Box 163
Mount Ida, AR 71957
(870) 867-3556 or (870) 867-3719 after 5:00 p.m.
http://www.mtidachamber.com/stanley/index.html

DIRECTIONS:
Stanley Rock Shop is located on Pine Street off U.S. Highway 270, east of Mount Ida.

SEASON:
Open all year round

HOURS:
8:30 a.m. to 5:00 p.m. daily

COST:
$10 per adult all day
$5 per child ages 5-16 all day
There is also a half day rate for those who arrive after lunch.

WHAT TO BRING:
Bring standard mining tools, your own food and plenty to drink. The mine owners recommend a hammer, screwdriver or tire iron. Anything you can scratch around with in the dirt will be fine. The owners kindly remind us that red clay stains are difficult to get out of clothing, so wear old clothes to the digging site.

INFORMATION:
The Fisher and Stanley families have mined this area through five generations. During World War II, the fine gem-quality crystals were used for oscillators (radio chips). The family rock shop has been in business since the 1940s and still occupies the same location. Over three hundred types of minerals are for sale at the shop. The crystal clusters from the mine are particularly decorative. The family's private collection has become a small museum which is available for viewing upon request.

The Stanley Mine is located 10 miles from the shop. You can drive all the way up Fisher Mountain to the digging site. There are lovely mountain views from the mine area. Enjoy the scenery while you try your luck.

You will need to clean the crystals that you find at the mine in a mild acid to remove dirt and rust stains. Use one pound of oxalic acid and two and a half gallons of water. Use any container but aluminum, and then heat the water until all the acid is dissolved. Place the crystals in the solution and cover. Allow the crystals to soak for four or five days. The solution will be good for up to three uses.

ADDRESS:
Route 1, Box 364
Murfreesboro, AR 71958
(870) 285-3113
http://www.craterofdiamondsstatepark.com/
craterofdiamonds@arkansas.com

DIRECTIONS:
The park is situated midway between Hot Springs and Texarkana, Arkansas. From Texarkana take Interstate Highway 30 north and exit at Hope (Exit 30). Take State Highway 4 northwest to Nashville. Then take State Highway 27 northwest towards Murfreesboro. At Murfreesboro take State Highway 301 southeast to the park.

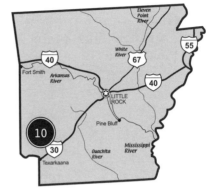

SEASON:
Open all year round

HOURS:
8:00 a.m. to 5:00 p.m.

COST:
$5.00 – Adults
$2.50 – Children

WHAT TO BRING:
Bring equipment for diamond mining. Also include a large washtub, sprayer bottle, rubber boots, and screens with varying size mesh (quarter- and one-eighth-inch). A plastic container with a lid is suggested for holding your finds. Equipment may also be rented from the park with a deposit and I.D.

INFORMATION:
   In this rare thirty-six acre field, diamonds of various colors and sizes can be found in their natural matrix. They are not easy to find, but an average of over six hundred diamonds are found here each year. Over 70,000 have been found so far. The largest recovered was the 40.23-carat diamond called the Uncle Sam. The 15.33-carat diamond, Star of Arkansas, worn by Mrs. Clinton at her husband's inauguration, was also found here.
   Diamonds can be located by surface inspection, surface scratching, sifting, digging and washing. Dirt is turned monthly to expose new soil. You are looking for well-rounded crystals usually smaller than a pea. Diamonds are naturally coated with oil and resist mud or dirt. Look for clean crystals. The most common colors are white, brown and yellow. Bring anything you think might be a diamond to the visitor center for verification and weighing. Anything you find, you keep.
   Diamonds are the chief attraction here, but don't overlook the other gems and minerals such as amethyst, agate, jasper, quartz, calcite and barite. The visitor center offers orientation programs on nature, geology, mining methods and history. The park has sixty campsites with water and electrical hookups. There are restrooms, exhibits, a bathhouse, and gift shop within the park. Restaurants and motels are nearby.

ADDRESS:
P.O. Box 8219
Hot Springs, AR 71910
(501) 984-5396 or (800) 291-4484
colemans@cswnet.com
http://www.colemanquartz.com

DIRECTIONS:
From Hot Springs take State Highway 7
northwest for 14 miles to the mine.

SEASON:
Open all year round

HOURS:
8:00 a.m. to 5:00 p.m.
8:00 a.m. to 6:00 p.m. – Summer

COST:
$20 – Adults
$5 – Children ages 7 to 16
Free – Children ages 6 and under

WHAT TO BRING:
Bring standard mining tools, your own food and plenty to drink. Tools are also
available for rent from the mine.

INFORMATION:
   This area has been mined commercially for many years. The ground at the mine is an open
pit, turned over by a bulldozer. Over forty acres of tailings are available. Tailings piles are piles
of stone broken up and searched by the mine. Many good crystals are overlooked. Crystals are
guaranteed to be found by experienced rockhounds or "pebble pups." The mine provides
crystal washing stations in the digging area.
   The mine has a gift shop and a wholesale showroom. The mine is located close to Hot
Springs National Park. After visiting the park, you might want to visit Coleman's Gift Shop and
browse their large selection of quartz, amethyst, agate and other minerals.
   The Crystal Ridge RV Park is located on the grounds of the mining company and has sites
for RVs and tent campers. Twenty-six sites include water and electricity. Modern restrooms,
dump stations and laundry are available to guests.

ADDRESS:
Willis Crystal Mine
21250 Buffalo Road
Paron, AR 72122
(501) 594-5228 or (501) 594-5589
willisquartz@aol.com
http://www.rockhoundingar.com/dealers/willis.html

DIRECTIONS:
The Willis Crystal Mine sits between Perryville and Crows, Arkansas off Route 9. From Interstate Highway 430 drive 25 miles south through Perryville to Williams Junction. Continue south on Route 9 go 5.9 miles to Brown's Corner and turn right onto Lake Winona Road. Drive 1 mile to the fork and turn left onto Buffalo Road. Drive half a mile to the mine entrance on your right.

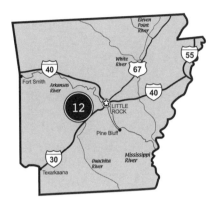

SEASON:
Open by appointment only

HOURS:
By appointment only

COST:
$5 – Digging the tailings
$10 – Digging on the veins
Free – Children ages 12 and under with a paying adult
Stay all day – keep what you find

WHAT TO BRING:
Bring small garden tools for digging, rock hammers, protective glasses and gloves. Some of these items are available at the mine. Wear old clothes, sunscreen and a hat.

INFORMATION:
The mine is an open pit on ten acres and you can drive your car to the mine site. This area is noted for the chlorite inclusions in quartz. Crystals of good size and unusual shapes are found here, including parallel growths and tabular crystals. Guests may dig on the vein or tailings.

The largest "points" found here was 330 pounds, unearthed in 1951, and a 750-pound cluster. So check the struts on your car and come for a visit.

Nearby you will find a covered picnic area and port-o-potty. Water is available at the gift shop, which is wheelchair accessible. Buses and large RVs are welcome.

ADDRESS:
Sweet Surrender Crystal Mine
60 Mary's Eagle Trail
Mount Ida, AR 71957
(870) 867-0104
randy_skates@earthlink.net
http://www.arcrystalmine.com/ssmine.html

DIRECTIONS:
From Mount Ida, take Highway 27 North approximately 10 miles to Horseshoe Bend Road (half a mile past Wishita on the right). Turn right and drive half a mile to the Forest Service Road. If you come to the first mailbox on the right on Horseshoe Bend Road you went too far). Follow the service road to the mine at the top of the hill.

SEASON:
You must call Randy Skates (870) 867-0104 to be sure the mine will be open.

HOURS:
Call for hours (see above)

COST:
$20 – Adults fee for digging
$10 – Children ages 6 to 10 fee for digging

WHAT TO BRING:
This is a primitive site with no facilities, services or telephones. You will need to bring all equipment, food and tools. Some of the tools you need should include: hand tools such as picks, long handled screwdriver and rock hammers. Bring a bucket or two, gloves and eye protection. A hat and sunscreen are advised along with newspaper to wrap your best crystals. Don't forget plenty to drink and something to satisfy hungry miners.

INFORMATION:
   Sweet Surrender Crystal Mine is a dig your own crystal mine inside a working commercial crystal mine. According to the Chamber of Commerce in Mount Ida, you have a better chance to find crystals in a fee dig mine that is currently in commercial operation. Sweet Surrender fits that bill. Crystals must be dug by hand to avoid damage and be aware that the red mud here covers everything and stains shoes and clothing. This is true at all the Arkansas crystal mines. If you want to get away from day-trippers and off the beaten track this mine may be right up your alley (or mountain).

MUSEUMS

Site 14.
Crater of Diamonds State Park
Route 1, Box 364
Murfreesboro, AR 71958
(870) 285-3113
http://www.craterofdiamondsstatepark.com/
Included in the park visitors' center is a small museum on the history of diamond mining in Arkansas. Some real diamonds and reproductions of the largest diamond ever found here are on display here.

Site 15.
University Museum
202 Museum
University of Arkansas
Fayetteville, AR 72701
(870) 972-2074
http://museum.astate.edu/exhibits.html
This university museum contains a collection of rocks, minerals, and fossils indigenous to Arkansas. Among the many mineral samples are quartz crystals, which is the official state mineral of Arkansas.

CAVES

Site 16.
Blanchard Springs Cavern
P.O. Box 1279
Mountain View, AR 72560
(888) 757-2246
http://www.fs.fed.us/oonf/ozark/recreation/caverns.html
Visitors take a two-mile tour along paved, lighted trails through a "living" cave with many formations still growing. This site includes an exhibition hall, movie, and visitor center.

Site 17.
Bull Shoal Caverns
P.O. Box 444
Bull Shoals, AR 72619
(800) 445-7177
http://www.bullshoalscaverns.com.
This Ozark cavern of limestone formed over 350 million years ago and is still growing. The cave has a gift shop and in located near Mountain Village, circa 1890.

Site 18.
Cosmic Caverns
6386 State Highway 21 North
Berryville, AR 72616
(870) 749-2298
http://www.estc.net/cosmiccavern/
This Ozark cave boasts a large underground lake, which appears to be bottomless and is home to blind rainbow trout. There is a gift shop on site.

CAVES (CONTINUED)

Site 19.
Hurricane River Cave
Highway 65
P.O. Box 240
Pindall, AR 72669
(800) 245-2282
http://www.hurricanerivercave.com
An ancient subterranean riverbed created channels cut by water. Many unique and beautiful formations grace these chambers. In addition to stalactites, soda straws and column formations, this cave contained the skeleton of saber-toothed cat, bear, and Native American Indian, who still remains in his final resting place. Picnic on site and enjoy the gift shop.

Site 20.
Mystic Caverns
Highway 7 South
P.O. Box 1301
Harrison, AR 72602
(888) 743-1739
http://www.mysticcaverns.com
Discovered in the 1850s, this Ozarks cave has many interesting formations including a domed cavern eight stories high. Guided tours are scheduled throughout the day.

Site 21.
Onyx Cave Park
388 Onyx Cave Lane
Eureka Springs, AR 72632
(501) 253-9321
This cave has many beautiful rock formations and a unique feature – electronic headphones for a self-guided tour, which means no waiting around for the next tour.

Site 22.
War Eagle Caverns
21494 Cavern Dr.
Rogers, AR 72756
(501) 789-2909
http://members.tripod.com/~WarEagleCavern//
This cave opened in the 1970s making it a new addition to existing show caves. The owners left the cave in its natural state, which includes the caves natural entrance, leading to an underground stream. Guided tours depart every forty minutes and take visitors past interesting formations and through large chambers. Fossils are in abundance. There is a gift shop on premises.

Deposits of iron became critically important to the nation's survival

Saugus Iron Works was the first foundry in North America opening in 1646. Several of the early settlers to Connecticut discovered sources of iron ore. These early furnaces used bog iron to create crude blister iron, which needed further refining to create wrought iron. Better quality iron ore was discovered in 1716. The surrounding forests provided the fuel to make charcoal and the stage was set for America's first iron making industry. Blast furnaces sprang up in northwestern Connecticut in Kent, Canaan, Norfolk and Salisbury. In the 1730's, smithies wrought tools for colonial America. These early smithies began to break the chain of dependence forged by England. In the 1740s the blast furnaces were remodeled to produce cast rather than wrought iron pieces and another important step toward independence was laid. Until this time cast iron pieces, ranging from pots to cannons all came from England. Now America could produce their own tools and more importantly – weapons.

When war broke out in 1776, these iron works factories became critically important to the survival of our new nation. Connecticut developed into America's leading cannon-making center due not only to the casting technology, but also to a discovery made by Daniel Bissell in 1731. He found a large, high-grade iron ore deposit near Salisbury, thus supplying the raw materials for arms. Connecticut provided a steady flow of iron ore, lead and limestone to the colonies. The new government considered this industry so vital, they exempted ironworkers from military service. Orders for cannons poured in to Connecticut and the plant worked frantically to meet demand. In total, Salisbury furnaces cast over 800 cannons of various sizes, grapeshot, rifle shot, hand grenades and of course, pots and pans. Without these products, the colonial coast would have been defenseless against British guns.

The smelting furnaces burned and iron fabricators of Connecticut continued their trade until the 1880s when resources ran low. Timber was no longer plentiful and ore resources dwindled. At the same time coal deposits were discovered in other states replacing charcoal as fuel for furnaces. Iron smelting moved west. Evidence of these early blast furnaces still dot the landscape of Connecticut in the form of glassy slag and blackened soil.

In addition to iron, copper, lead and granite were also mined in Connecticut. One site known for copper is Newgate. In 1705 this was the first chartered mine in the country. After thirty years, it failed to turn a profit. The state took over and decided to use the mine to hold criminals. In 1776 Newgate became America's first prison. With typical New England frugality in mind, it was thought that the cost of prisoner's confinement could be deferred by having the men work the mine. Convicts were incarcerated in the mine, but there were no fences, so escapes were rampant. In 1777 the guardhouse burned to the ground and the facility was abandoned until 1780 at which point another attempt was made to make prisons profitable. The prisoners made nails and mined copper, but no profit was realized. Coopering, cabinetmaking, shoemaking and a wagon shop were all added along with a man-powered grain grinding treadmill. All efforts failed and the mine was abandoned in 1827.

Another Connecticut endeavor was more successful. In the area from Portland to Middletown prehistoric sediments collected into a rock known as brownstone. Brownstone was soft enough to carve, and polish and was easily quarried. In 1690 James Stanclift built the first house made of brownstone in what is now Portland. He also carved local gravestone of this material. By 1866 many churches, schools and homes where made from rock quarried here. The town employed over 1500 workers. Twenty-five ships took the brownstone down river to New York and up the Atlantic to Boston. Some stone traveled as far as London and San Francisco. Homes made of this material became known simply as brownstones. But all things must end and by the 1890s concrete production signaled the demise of this industry. In 1936 the quarry flooded and the quarrying ceased.

Other stones were quarried in this state including: marble, mica, feldspar and clay. Open pit marble was mined from 1734 to 1923. Barite, mined in Cheshire beginning in the 1800s, is today used in paint, glaze and as an additive to drilling oil.

Connecticut has other riches. Tourmaline is found here as well as rose quartz and garnets, which are the official state stone. Garnets found in Roxbury were mined for abrasives. The specimens unearthed here are not considered gem quality. Rockhounds can mine their own garnets at the Green's Garnet mine in Southeastern Connecticut.

One of the most unique gems of Connecticut is danburite. This clear crystal was first discovered in 1839 and was named for its place of origin, Danbury, Connecticut. Danburite is hard enough to cut and set in jewelry, but is little known. Though usually clear, danburite does occur in yellow, brown and pink. Currently this gem also is found in Caracas, Mexico, Japan and Burma but no longer in Danbury. Unfortunately the area where it was found is now buried beneath the city.

But construction work does not always bury treasures, sometimes work reveals them. This was just the case in 1966 when a construction worker unearthed one of this states greatest paleontological discoveries. The bulldozer operator exposed a slab of bedrock upon which he noticed several footprints. Work ceased and the site was protected. There are over 2000 footprints dating from the Jurassic period at this site, making it the largest known dinosaur trackway in North America. Most tracks are twelve-inches long indicating a carnivore of at least twenty feet in length and over 1,000 pounds. These impressions, left in soft sand, are now preserved at Connecticut Dinosaur State Park in the exhibit area. The park is open to all those who wish to stare back into the past and wonder.

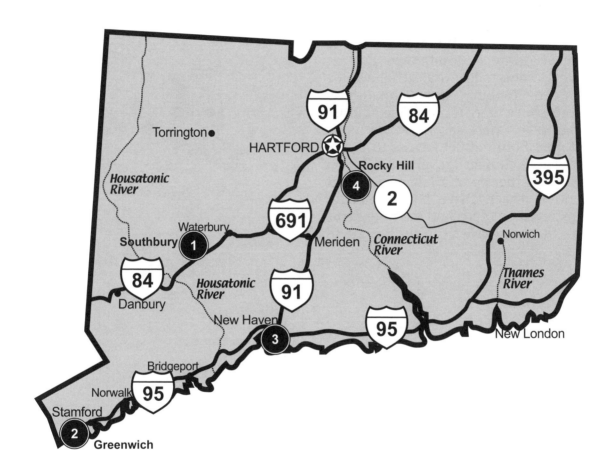

Diggings
1. Green's Garnet Farm–Garnets
Museums
2. Bruce Museum.
3. Peabody Museum of Natural History
Points of Interest
4. Dinosaur State Park

ADDRESS:
Perkins Road
Southbury, CT 06488
(203) 264-3550 (call in the evenings)

DIRECTIONS:
Green's Garnet Farm is in Southbury, Connecticut, west of Waterbury. From Interstate Highway 84 east, take Exit 15 (Southbury). Then follow State Highway 67 northwest to Roxbury, approximately 8 miles. In Roxbury there is a triangle of grass with a flag pole where several roads meet. Turn left onto South Street, across from the white Episcopal church. Travel 4.4 miles on South Street to the top of a hill. Make a left onto Perkins Street. Drive half a mile to where the pavement ends and the road turns sharply to the left. Drive through the gate to the house.

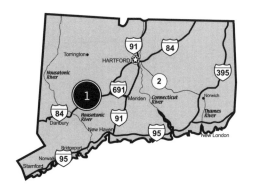

SEASON:
Open all year round

HOURS:
9:00 a.m. to 5:00 p.m.

COST:
$2 per car – Fee for parking
Free – Fee for collecting

WHAT TO BRING:
Bring standard mining tools, your own food and plenty to drink.

INFORMATION:
    The garnets found here are well-formed, dodecahedron crystals of a dark red or purple color. Some garnets are of gem quality and can be faceted into jewelry. The majority are dark and opaque.
    Garnets are plentiful at the farm, making it a favorite spot for local rockhounds. The Archibalds are very casual about allowing collectors onto their property. You can collect all day and there is no limit on the number of garnets you carry out.
    The terrain is rocky and there are some cliffs and lots of holes; watch your step. Fill in all holes you make, and carry out all your trash. Garnets can be found by raking or screening through loose dirt. Many stones are exposed through the weathering process of the host rock. The more ambitious miner can break large rocks or cliff face to uncover garnets. This involves climbing and using a sledgehammer and chisels.

Museums

Site 2.
Bruce Museum
One Museum Drive
Greenwich, CT 06830
(203) 869 0376
http://www.brucemuseum.org/
The geology collection includes outstanding aesthetic specimens of minerals, rock and fossils.

Site 3.
Peabody Museum of Natural History
Yale University
P.O. Box 208118
170 Whitney Avenue
New Haven, CT 06520-8118
(203) 432-5050
The mineralogy division maintains a collection of over 35,000 specimens of rocks and minerals including a collection of gemstones, and a large collection of Connecticut minerals.

Points of Interest

Site 4.
Dinosaur State Park
400 West Street
Rocky Hill, CT 06067-3506
(860) 257-7601/ (860) 529-8423
http://www.dinosaurstatepark.org/
Come visit one of the largest dinosaur track sites in North America protected beneath a giant dome. Also within the dome are life-size Triassic and Jurassic dioramas and interactive displays. These Jurassic fossils are 200 million years old. The park also has nature trails and an arboretum containing plant species that grew when dinosaurs walked the earth.

Light a flare, soothe your stomach and build strong bones all with one versatile element.

Delaware ranks 49th in the country in square miles, kept from last place only by the state of Rhode Island. This state is only 96 miles long and in places only nine miles wide. Needless to say it is not awash in mineral riches. You might think that such a small state would have no mining activity, but it does. This state takes advantage of what it has. Most commercial mining involves sand, gravel and some magnesium compounds.

Extracting magnesium may better be called producing rather than mining. This mineral is abundant, ranking as the eighth most common element in the earth's crust. Seventy percent of United States magnesium compounds come from seawater or brine water and Delaware is one of three largest producing states. The others two are Florida and California.

The process for extraction involves electrolysis, which caused the magnesium to fuse with chloride in the water as magnesium hydroxide. This is treated with lime from crushed oyster shells. The hydroxide is collected and treated to form chloride and the magnesium precipitates out. One pound of magnesium can be collected from each 100 gallons of seawater, providing a nearly inexhaustible source of this metal.

Magnesium is a tough, silver-white metal that burns with a bright white flame, making it ideal for use in in flares, pyrotechnics and flash photography. Mixed as alloy with aluminum it strengthens the metal in aircraft and car engines. During WWII, Delaware's magnesium production became vital to the war effort. Magnesium has more mundane uses as well. Mixed with salts magnesium becomes Milk of Magnesia to soothe your stomach. Mixed with another type of salt and you have Epson salts to soak your tired feet. In 1765 a professor named Hilderbrandt discovered magnesia in human bones. If you check your multiple vitamins you should find trace amounts of magnesium there.

Speaking of bones, the area around the Chesapeake and Delaware canals is the site of most of this state's fossil discoveries. These fossils from the Cretaceous period are mainly marine animals including a duck-billed dinosaur, giant crocodile and many types of bony fish, turtles and fossilized seashells. Collectors and paleontologists have had good luck searching for these treasures in soil removed from the canal area.

Delaware also has some very interesting minerals. Dolomite, epidote and graphite are all indigenous. Precious and semiprecious stones include: rutile quartz, staurolite, labradorite, beryl and spinel.

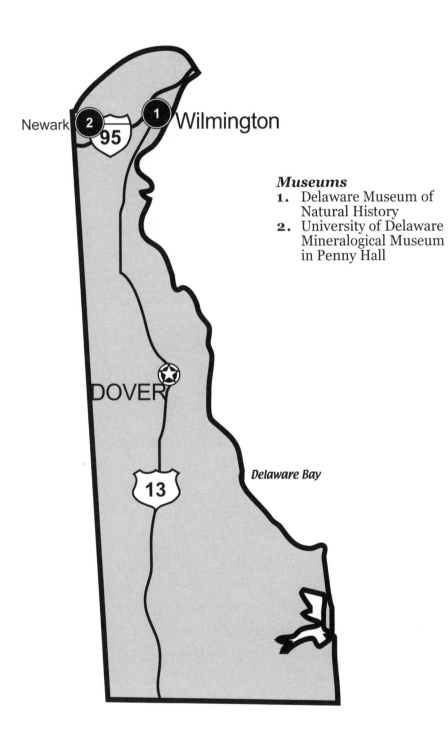

Museums
1. Delaware Museum of Natural History
2. University of Delaware Mineralogical Museum in Penny Hall

MUSEUMS

Site 1.
Delaware Museum of Natural History
4840 Kennett Pike
P.O. Box 3937
Wilmington, DE 19807-0937
(302) 658-9111
http://www.delmnh.org/
Most people have a collection of seashells, but this place has nearly a quarter-million of them! This museum also has a collection of 117,000 birds. Don't forget to visit the discovery room for kids.

Site 2.
University of Delaware Mineralogical Museum in Penny Hall
114 old College
Newark, DE 19716
(302) 831-4940
http://www.udel.edu/geology/min/
The Irénée du Pont Mineral Room exhibits over 750 mineral samples including gems and carvings. The teaching collection includes 6,000 specimens. Be sure to see the crystalline gold and lovely red rhodochrosite.

Spanish Galleons and Silver and Gold

Spaniards called the coral reef, surrounding much of Florida's coast "the dragon's teeth" with good reason. These spiny living walls ripped the bottom out of more than one ship caught in a storm. The Spanish were not the only one's tossed upon the reefs. Countless ships have sunk off the coast of Florida through the centuries. So many ships were lost that "Wrecking," which is the salvaging of cargo from stranded and sinking vessels, became a very profitable business in Key West in the 1800s. Recovering lost treasure is a hobby to some and a life's work to others. Treasure salvagers hunt the coastal waters each year to bring fabulous finds to the surface.

Among the most famous discovered wrecks in Florida waters are the 1715 Fleet and Key West's Atocha. In the 1960s, Kip Wagner, a scuba diver and adventurer discovered blackened discs upon the shore near Sebastian Inlet and Vero Beach. Assuming they were bits of metal from wrecked hulls of ships lost over the years upon the reefs, locals would often skip them across the water. They were partly right. They were from a shipwreck on the reef and they were metal. But rather than the rusting bits of iron they suspected, Kip learned they were Spanish pieces of silver – real pieces of eight. After some research he discovered that a fleet of twelve Spanish ships ran into a hurricane in 1715 and wrecked on the savage Florida coast. Salvage efforts began immediately and continue to this day. There is a very interesting museum called McLarty Treasure Museum, which explains all about the wrecks, early salvage attempts by the Spanish and the plight of the survivors. To see some of the riches brought from the sea visit Mel Fisher's Treasure Museum in Sebastian Inlet. One of the most impressive recent finds was discovered in 1993: a diamond studded brooch, necklace and bob earrings. Metal detector hobbyists flock to beaches on the treasure coast with good reason. A gold dragon used as toothpick and whistle was found on the beach at Sebastion Inlet. Hobbyists need to know the salvage laws and stay only on the beach. Detecting in the ocean on a salvage claim is illegal.

One of the early treasure hunters of the gold coast was a former chicken farmer and scuba shop owner named Mel Fisher. The treasure bug bit hard and he began a search for a wreck called the Nuestra Señora de Atocha, which sank in 1622. He and his crew searched for sixteen years finding only small artifacts. He grew famous for telling his crew, "Today's the day" and this became the rallying call for his divers. On July 20, 1985 he found his ship. Gold, emeralds, silver bars, gold chains, jewelry and two fabulous emerald encrusted gold crosses have been uncovered to date. The recovery efforts are ongoing. Those with extra cash can buy stock in Mel's treasure salvage company, which entitles shareholders to a dive on the wreck. The Mel Fisher Maritime Heritage Society and Museum in Key West is a treasure trove of history and artifacts.

For some people fossils are a treasure worth hunting. Fossil hunters will love Venice Beach and Casperson Beach on the Gulf Coast. Many people come to these locations to find sharks teeth. They occur in great variety and color. Most seekers use some kind of a sorting screen or a device called a Florida snow shovel to sift sand. Scuba divers are lucky to search deeper water for larger teeth. But other places in Florida also are known for fossils and inland excursions to rivers and streams are popular outings.

Florida is known for phosphate mining. This essential ingredient in fertilizer is found in an area between Lakeland and Charlotte Harbor. The area is called Bone Valley, because of the ancient bone remains of the Miocene period unearthed here, including giant sloths, mammoths, mastodons, horses and bison. Some of the finds discovered while mining are located in the Bone Valley Museum. Many divers search Florida's lakes and rivers for artifacts of ancient people and fossilized bones. Diving for fossils requires a permit obtained from the state of Florida. For activities that don't require a permit or a cash investment, visit Captiva and Sanibel Island, one of the best sites in Florida to hunt for seashells. Metal detector hobbyists will love finding modern treasure in the form of jewelry and coins left by more recent visitors to Florida's beaches and oceans.

Spanish silver and remnants of an ancient world, Florida offers more than theme parks and white sands to those in search of treasure.

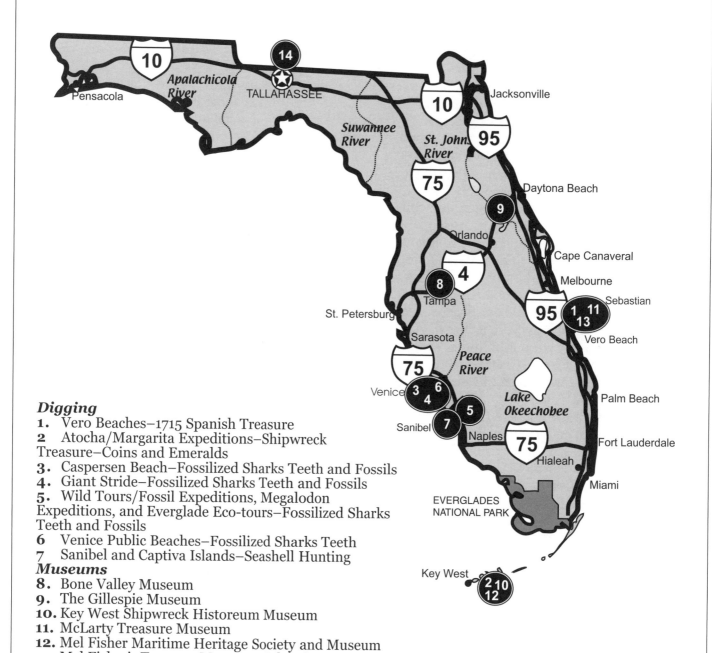

Digging
1. Vero Beaches–1715 Spanish Treasure
2 Atocha/Margarita Expeditions–Shipwreck Treasure–Coins and Emeralds
3. Caspersen Beach–Fossilized Sharks Teeth and Fossils
4. Giant Stride–Fossilized Sharks Teeth and Fossils
5. Wild Tours/Fossil Expeditions, Megalodon Expeditions, and Everglade Eco-tours–Fossilized Sharks Teeth and Fossils
6 Venice Public Beaches–Fossilized Sharks Teeth
7 Sanibel and Captiva Islands–Seashell Hunting

Museums
8. Bone Valley Museum
9. The Gillespie Museum
10. Key West Shipwreck Historeum Museum
11. McLarty Treasure Museum
12. Mel Fisher Maritime Heritage Society and Museum
13. Mel Fisher's Treasure Museum–Sebastian
14. Museum of Florida History

ADDRESS:
Highway A1A
Sebastian Inlet State Park, FL
(772) 589-2147
http://208.234.21.196/mclarty.htm

DIRECTIONS:
All beach sites are off State Highway A1A near Vero Beach, from Sebastian Inlet to Fort Pierce Inlet. Some beach accesses are: Wabasso Beach, Seagrape Beach and Turtle Trail Beach.

SEASON:
Open all year round. After winter storms are the most productive times for finding treasure.

HOURS:
See public beach signs for hours
(beaches are generally open from 7:00 a.m. to dusk)

COST:
$1 – Museum fee
Free – Beach access

WHAT TO BRING:
A good metal detector with a discriminator to exclude false signals from mineralized sand is essential. Also needed are a sand scoop and pouch to carry finds and remove trash. The shipwreck survivors were driven insane by mosquitoes, so bring bug spray and a hat.

INFORMATION:
On July 24, 1715, eleven Spanish Galleons were struck by a hurricane. Since that time gold and silver coins called escudos and reales have been found on these beaches. Lucky rockhounds have even found green emeralds from these ships. The beaches are good places for modern coins and shells. Please follow local regulations and stay on the walkovers to the beaches. Stay off the dunes, watch out for turtle nests and carry out your trash.

Use of a good metal detector is strongly recommended. Many ships, including metal ones, have broken apart on these beaches. So a good discriminator on your detector in a must. Coins are difficult, but not impossible to find. Using a metal detector in the water is forbidden by Admiralty Claims. Stay off private property. Stop and see the McLarty Museum 1.9 miles north of Sebastian Inlet (10:00 a.m. to 4:45 p.m.) when you need a break.

ADDRESS:
Mel Fisher Maritime Heritage Society Museum
200 Greene Street
Key West, FL 33040
(305) 296-6533 or (888) 577-6995
http://www.melfisher.com

DIRECTIONS:
Expeditions depart from Key West. Visit the museum office to make arrangements.

SEASON:
Diving seasonal and weather dependent so call to make a reservation.

HOURS:
One day – two dives

COST:
$6,250.00

WHAT TO BRING:
Bring your diving gear, scuba diving certification card and log book. Tank rentals are available in Key West.

INFORMATION:
    Certified divers with some big money to spare can arrange to dive on a real Spanish treasure ship. Divers do NOT get to keep what they find. Your $6250 fee actually buys you a one-eighth unit share in this salvage operation and as a benefit you can arrange a one day dive on either the Atocha or Margarita site. These ships sank in 1622 in a hurricane and have yielded over $400 million in treasure to date.

    Divers must provide their own transportation to the salvage ship. That can be arranged with charter ships in Key West.

    At the end of the dive season there is a distribution of treasure to shareholders. Depending on your investment and the year's haul you receive your portion paid in coins, emeralds or other salvaged treasure. What an exciting way to own a piece of history!

ADDRESS:
Venice Area Chamber of Commerce
257 Tamiami Trail North
Venice, FL 34285
(941) 488-2236
http://www.venice-fla.com/

DIRECTIONS:
South of Venice and the airport on Harbor Drive.

SEASON:
Open year round

HOURS:
Daylight

COST:
Free

WHAT TO BRING:
Swimsuit, umbrella, sunscreen, "Florida snow shovel" or other sifting screen, aquaboots, mask and fins, container for finds. In the winter a light wetsuit is advisable for those who want to search in the water. Scuba divers may make a beach dive from this site.

INFORMATION:
The longest beach in Sarasota County is known for its beauty, seclusion, inlet nature trails and fossils. The most common fossils found are sharks teeth, but other treasures are waiting, including stingray tooth plate and barbs, fossilized shells, bones from deer and other mammals, turtle shell plates, vertebra from various species. Fossils are generally brown, black or gray depending on the material in which they rested. Fossils found on the beach are usually between a quarter-inch to two inches in length. Divers, or those lucky enough to search after turbulent weather, have a greater chance of unearthing larger specimens. Once you start collecting teeth you'll need to identify the species. Check out the stores in downtown Venice for books, fossils and other goodies.

ADDRESS:
509 Tamiami Trail
Venice, Florida 34231
(914) 922-3483/(941) 809-2689
http://www.fdu.com/sharktooth.htm

DIRECTIONS:
Located in Fisherman's Wharf Marina in Venice off U.S. Highway 41. From Interstate Highway 75 take Exit 37 (Clark Road) and go west until you reach U.S. Highway 41, then turn south on U.S. Highway 41 and drive until you reach the split. Bear right and drive to the marina.

SEASON:
Open year round

HOURS:
Two daily charters, morning and afternoon (three minimum to run) with a maximum of six divers per trip.

COST:
Collection Dive: $45 for Snorklers
$55 for one tank dive, $65 two-tank dive for scuba divers

WHAT TO BRING:
Scuba gear, certification card, snorkel equipment, collecting bag, personal gear, hand rake.

INFORMATION:
Diving or snorkeling off Venice, Florida is a great way to find shark teeth. Divers searching for rare six-inch specimens can try their luck with Giant Strides who offer daily collecting charters. Average dive depth is only fifteen feet for snorkelers and thirty for scuba divers. Bring your own equipment or rent for an additional fee.

ADDRESS:
213 Lincoln Avenue
Lehigh Acres, FL 33972
(800) 304-9432/(239) 368-3252
fossilx@earthlink.net
http://www.fossilexpedition.com/main.htm

DIRECTIONS:
Call for directions

SEASON:
Year round. Be advised that the Everglades area temperatures are more pleasant, with fewer bugs in the winter months. Megalodon expeditions depart twice a week.

HOURS:
Hours vary depending on the type of trip.
9:00 a.m. to 3:00 p.m. is the average day.

COST:
Call for prices, which vary depending on the type of trip.

WHAT TO BRING:
Shallow River Expeditions - Bug spray, a change of clothes including shoes, hat, gloves, hand tools, sifting screens.

Quarry Expeditions - hat, sunscreen, old clothes, tennis shoes and socks, a cooler with lunch and water, aluminum foil (to wrap fragile finds), nylon collecting bag, zip-lock bags and screwdriver or other digging device.

INFORMATION:
These excursions take place in Tampa, Saint Petersburg, Orlando, Sarasota, Naples and Fort Lauderdale. Expeditions are generally conducted in shallow streams and rivers. Group members can snorkel or stand in shallow water and screen-wash material in search of fossils. Florida has more than a thousand streams and rivers in which to search, so opportunities are many and varied.

Mammoths, mastodons, camels, sloths, horses, tortoises, saber-toothed cats, tapirs, jaguars, giant armadillos, whales, sharks and dolphins all lived in southern Florida, leaving behind evidence of their passing.

Expedition leader and owner Mark Renz also leads expeditions for fossilized seashells in area quarries.

ADDRESS:
Chamber of Commerce
257 North Tamiami Trail
Venice, FL 34285
(941) 488-2236
http://www.venice-fla.com/vacc.htm

DIRECTIONS:
Take Interstate Highway 75 or U.S. Highway 41 to Venice, Florida. Take Venice Avenue to the Venice Public Beach.

SEASON:
Open all year round

HOURS:
Daylight

COST:
Free

WHAT TO BRING:
You can find teeth without equipment. Your success rate will be better if you have a metal tooth scooper with a long handle, homemade screens or even pasta strainers. Bring a closed container to keep your shark's teeth. Swimsuit pockets tend to give back your treasures to the sea, and are not recommended.

INFORMATION:
    All of Venice's public beaches are good places to look for shark teeth. Try Venice Beach, Venice Fishing Pier and Caspersen State Park. Many teeth are found washed up on the beach or in ankle-deep water in the surf. Some enterprising individuals snorkel or scuba dive for teeth and other fossils. Look for black sand and small rocks, these are fossil deposits. People on the beach are friendly and willing to share hunting techniques with you. They will also show you their haul if asked.

    Teeth range in size from a quarter-inch to six inches. You will find more of the former and fewer of the latter. Bring sandals or sneakers. The black sand gets HOT and you'll never make it back to your car without them. Keep your eyes open for fossilized bones which are black with little holes in them where blood vessels once were. Stop by the Sea Pleasure and Treasures Gift Shop in Venice and see the amazing teeth on display and for sale.

ADDRESS:
Sanibel & Captive Islands Chamber of Commerce
1159 Causeway Road
Sanibel, FL 33957
(239) 472-1080

DIRECTIONS:
State Highway 867 off of U.S. Highway 41

SEASON:
Open all year round

HOURS:
Daylight or night (best finds occur at the changing of the tides)

COST:
Free

WHAT TO BRING:
You may want to bring a pail, shell identification book, flashlight for night hunting and a really loud tropical shirt.

INFORMATION:
Sanibel and Captiva Islands are renowned as the best shell hunting beaches in the world, and Turner Beach is known for the best collecting. Keep in mind there are so many great shells to be found here, that you should not take live shells. This will destroy future populations, so you should return shellfish that still reside in the shell you may pick up.

The best time to search is early in the morning when the tide is going out. Bring a flashlight for searching at night as well. Early risers are rewarded by having the days first pick of scattered shells.

Most of Florida's barrier islands have some fine shells, but Sanibel and Captiva are unique. Barrier islands lie parallel to the coast, but Sanibel is shaped like a fish hook. The portion of beach that sticks out into the gulf acts to catch shells.

The community here has done an excellent job of controlling development. The building codes are very strict. For you, this means that there are no high-rises spoiling the view. From the beach, the hotels and condos are all at tree level and set back, to be unobtrusive. The shops and stores are also low and recessed with small, non-illuminated signs, creating a very picturesque spot for a vacation.

MUSEUMS

Site 8.
Bone Valley Museum
State Highway 37
Mulberry, FL 33860
(863) 425-2823
http://www.mulberrychamber.org/phosmuseum.asp
Fossilized remains of prehistoric animals are common in phosphate deposits. This museum displays a collection of these fossils, area memorabilia and exhibits on the local phosphate mining industry.

Site 9.
The Gillespie Museum
Stetson University
Unit 8403
DeLand, FL 32720
(386) 822-7330
http://www.gillespiemuseum.stetson.edu/
Florida's mineralogy, geology and paleontology are the focus of the Gillespie Museum, which houses 25,000 mineral specimens including examples of Florida's minerals, fluorescent minerals in a special dark room, fossils (including a mosasaurus dinosaur), precious and semi-precious stones and crystalline minerals.

Site 10.
Key West Shipwreck Historeum Museum
1 Whitehead Street
Key West, FL 33040
(305) 292-8990
http://www.historeum.com/
This is a combination of wrecker's museum and reenactment. Shipwreck recovery was once a central industry to this island's economy and wrecking continues today. This museum combines film, lasers, live actors and traditional museum displays to make history come alive. Also included in the admission price is a 60-foot lookout tower once used to spot wrecks on the reefs and now providing a spectacular view.

Site 11.
McLarty Treasure Museum
13180 North State Road A1A
Sebastian, FL 32958
(772) 589-2147
http://www.atocha1622.com/mclarty.htm
Located at the site where the 1,500 survivors from the shipwrecked 1715 fleet made camp, this museum offers exhibits of items from the wrecks including gold, silver, jewelry and porcelain along with historical items and a description of the disaster.

Site 12.
Mel Fisher Maritime Heritage Society and Museum
200 Greene Street
Key West, FL 33040
(305) 294-2633
http://www.melfisher.org/
Fabulous treasure troves await you at this customhouse turned museum. Artifacts from the 1622 Spanish shipwrecks the Nuestra Señora de Atocha and Santa Margarita as well as exhibits on the salvage efforts fill the building. Visitors can lift a silver bar, see gold and silver bars and coins, jewelry (including a six-inch gold cross studded with emeralds) and other items salvaged from the ships.

Site 13.
Mel Fisher's Treasure Museum –Sebastian
1322 U.S. Highway 1
Sebastian, FL 32958
(561) 589-9875
http://www.melfisher.com
This museum displays treasure recovered from the coasts of Florida. Much of the treasure here comes from the 1715 Fleet of twelve ships that wrecked on "the dragon's teeth" as the Spaniards called Florida's jagged coral reefs.

Site 14.
Museum of Florida History
R. A. Gray Building
500 South Bronough Street
Tallahassee, FL 32399-0250
(850) 245-6400
http://dhr.dos.state.fl.us/museum/
This museum serves as the repository for the state's historical artifacts and maintains permanent collections of nearly 44,000 items including the Silver Plate Fleet exhibit of artifacts, gold and silver recovered from the 1715 and 1733 Fleets.

The 1832 Gold Lottery and the Trail of Tears

Georgia produces more kaolin than any other state in the union and is a major source of barite, iron oxide pigments, feldspar and clay. What is kaolin? You probably use it every day. Kaolin is also called china clay and is white alumina-silicate with many applications. Eighty percent of the kaolin mined in Georgia is used in creating paper products. Did you ever wonder what gives glossy paper its shine? This material also makes house paint resist mildew and is used as a wire insulator, in cosmetics, tires, fiberglass, nylon and to make car and truck parts.

Georgia also has oil, gas, coal and peat. There are some semiprecious stones in this state including agates, onyx, obsidian and jasper. Metal resources include copper, manganese, lead, zinc, silver and gold. This last metal sparked some of Georgia's darkest moments in history.

Gold was known to exist by Native Americans, specifically the Cherokee tribes. In the 1540s a Spanish explorer name Hernando De Soto arrived in what is now Georgia and began mining along the Chattahoochee River. The Spanish continued their gold recovery until they were forced out of the territory. In the 1800s some early white settlers mined for gold on Cherokee land. By the 1830s news of the discovery of gold hit the country and Europe and the rush was on. Men came from every state and set up mining camps on Cherokee land. Between 1830 and 1837 the Philadelphia mint received over 1.7 million dollars worth of gold from Georgia. The Cherokee, who had lived peacefully on this land for many generations, now found themselves overrun with miners.

In 1830 the United States Congress passed the Indian Removal Act. Camps and forts were built and the Cherokee were herded off their land. The Cherokee Nation brought their suit to the Supreme Court who refused to hear the case. The building of forts was speeded and more Cherokee were removed from their land. In 1832 the state of Georgia held a land lottery, raffling off 40-acres gold-bearing lots to white settlers. These same settlers were stunned when the Supreme Court ruled in favor of the Cherokee and stated that Georgia could not extend its laws on a sovereign nation such as the Cherokee. Many early settlers in Georgia who had lived peacefully with the Cherokee agreed with the court. Most of the newer settlers, who now occupied their land, did not. In 1838 several members of a treaty party signed The Treaty of New Echota, effectively signing away the rights of all Cherokee and giving President Andrew Jackson the legal right to remove the Indians. In May of 1838 the Georgia Guard rounded up the Cherokee under orders of U.S. General Winfield Scott. In June of 1838 the Cherokee were sent on a forced march that came to be known as the Trail of Tears. Four thousand men, women and children died during the removal to Oklahoma.

That same year the placer gold grew hard to find. In 1849 the California gold rush effectively ended the Georgia Gold Rush, but not before destroying the Cherokee Nation.

One only has to look at Georgia's state capital in Atlanta to be reminded of this states gold rush past. The citizens of Dahlonega provided the sixty ounces of gold needed to cover the capital dome.

Those interested in learning more about Georgia's history could visit the Dahlonega Gold Museum in Dahlonega. Gold prospecting opportunities in Georgia include five gold mining areas: Consolidated Gold Mine, Crisson Gold Mine, Gold 'N Gem Grubbin', Hidden Valley Campground and Pine Mountain Gold Mine.

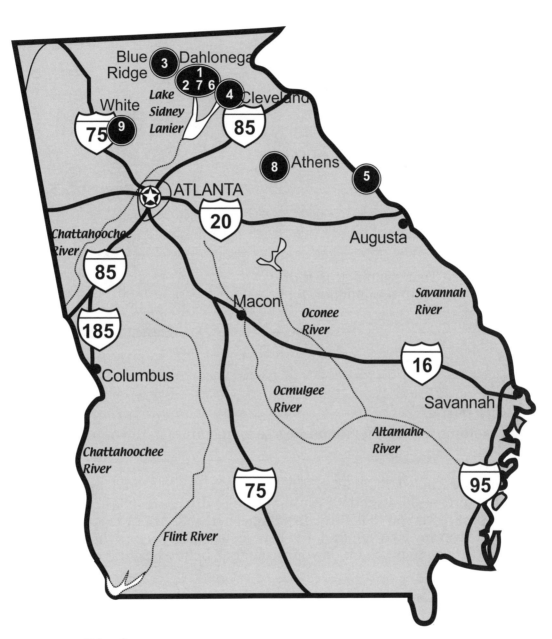

Diggings
1. Consolidated Gold Mine–Gold
2. Crisson Gold Mine–Gold Panning
3. Hackney Farm–Staurolite
4. Gold 'N Gem Grubbin'–Gold and Gems
5. Graves Mountain–Gold Panning
6. Hidden Valley Campground–Gold Panning

Museums
7. Dahlonega Courthouse Museum
8. Georgia Museum of Natural History
9. William Weinman Mineral Museum

ADDRESS:
185 Consolidated Gold Mine Road
Dahlonega, GA 30533
(706) 864-8473
info@consolidatedgoldmine.com
http://www.consolidatedgoldmine.com/

DIRECTIONS:
Head east from Dahlonega on Highway 52. At the light turn right onto Consolidated Gold Mine Road.

SEASON:
Open year round

HOURS:
Winter: 10 a.m. Last Tour around 4:00 p.m.
Summer:10 a.m. Last Tour around 5:00 p.m.

COST:
Mine Tour and pan for gold.
Adults - $8
Children - $5

WHAT TO BRING:
Comfortable clothing appropriate for 60-degree temperature in the mine shaft.

INFORMATION:
   This 45-minute guided tour takes visitors into a Georgia gold mine, as it would have appeared at the turn-of-the-twentieth-century. Guides discuss displays, mining equipment and the geology of the area as you walk through a tunnel network. After the tour take a free gold panning lesson and keep any gold you find. For experienced prospectors the mine has a high grade panning set-up for you to brush up on your technique before heading out to an area stream to try your luck.

ADDRESS:
2736 Morrison Moore Parkway East
Dahlonega, GA 30533
(706) 864-6363
crissongoldmin@alltel.net
http://www.members.tripod.com/~goldpanner/crisson.html

DIRECTIONS:
The Crisson Gold Mine is located 2.5 miles
north of Dahlonega on U.S. Highway 19. Look
for the green and gold sign.

SEASON:
Open all year round

HOURS:
10:00 a.m. to 6:00 p.m. – Summer
10:00 a.m. to 5:00 p.m. – Winter

COST:
$2.50 – Panning
By the bucket:
$6.25 – Fee for two and a half-gallon ore
$9 – Fee for five-gallon ore
$4 – Fee for two-gallon gemstones
$7 – Fee for five-gallon gemstones

WHAT TO BRING:
The mine will supply you with gold pans and ore. For $.50 you can buy a small bottle to carry
your gold away.

INFORMATION:
The Crisson Gold Mine was established in 1847 during the Dahlonega Gold Rush and is
owned and operated by fourth generation gold miners. The mine has been open to the public
since 1970. This family-owned operation is now open to your family for fun and adventure.

The people at the mine will instruct you and help you pan or screen out your material. The
mine has an indoor facility for year-round treasure hunting. If you don't want to pan for gold
you can screen for gemstones. You may find rubies, emeralds and numerous other stones. The
Crisson Gold Mine has a gift shop open seven days a week. Gold, gold jewelry and gold nuggets
can be purchased. Some gold mining equipment used commercially is on display.

ADDRESS:
Hackney Farm
652 Bullen Gap Road
Blue Ridge, Georgia 30513
(706) 492-5030
jahw@bellsouth.net
http://www.gamineral.org/commercial-hackney.htm

DIRECTIONS: Drive north on Interstate Highway 575/SR 515 to Blue Ridge, GA. Pass the Blue Ridge sign. After the overpass, turn left onto Bullen Gap Road. Drive 7/10 of a mile bearing right until you reach a mailbox number 652. Take the gravel road on the right and head down the hill to the collecting area.

SEASON: All year round – please call in advance

HOURS: Daylight

COST:
Adults - $10 per day
Children - $2 per day

WHAT TO BRING:
Pick and shovel, buckets, 1/2 inch and 1/4 inch shifting screens, a container to hold your staurolite crystals.
Bring food and water along.

INFORMATION:
Staurolite, also known as fairy stones or fairy crosses, is the state mineral of Georgia with good reason. These fascinating iron oxide crystals are beautiful and full of local folklore. Carrying this crystal is considered good luck and you should have good luck digging for them in the new ditches opened at this mine. The occurring crystals here are small. Most specimens are single blades, intersecting crosses and occasionally a angular Maltese cross. Other minerals occurring here include hornblende, diorite and some gold ore.

There is a stream on premises to aid in screening and plenty of shade trees when you are ready for a rest.

If there is no one around, please leave your digging fee under the front door of the old cabin located on the top of the hill at the end of the dirt road.

ADDRESS:
Gold 'N Gem Grubbin'
75 Gold Nugget Lane
Cleveland, GA 30528
(706) 865-5454
goldngem@linkamerica.net
http://www.goldngem.com/

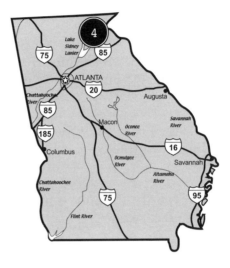

DIRECTIONS:
Drive 2 miles west from Cleveland Square, Georgia, turn right onto Town Creek Road and go 2 miles to the mine.

SEASON:
Open all year round

HOURS:
Winter: 10 a.m. Last Tour around 4:00 p.m.
Summer:10 a.m. Last Tour around 5:00 p.m.

COST:
Costs vary for panning and gem hunting.
GOLD:
$8 – Fee to pan with a standard one-gallon bucket for gold
$50 – Fee to pan with a one-gallon nugget bucket, guaranteed to yield a nugget
$95 – Fee for a two-gallon gold nugget bucket, guaranteed to yield nuggets
$19 – Fee to mine at the creek all day from the dirt brought from the mine
GEMS: (ruby, sapphire, emerald, topaz and amethyst)
$5 – Fee for a one-gallon standard bucket
$30 to $100 – Fee for a one- or two-gallon super bucket, price includes a discount of faceting and setting your find

WHAT TO BRING:
Standard equipment to pan and gem hunt is available for free. If you plan to mine by the stream, bring a lawn chair or something to sit on and a lunch. Dredges and highbankers are permitted and some are available to rent.

INFORMATION:
This hundred-acre site is unique for two reasons. It is in famous Dahlonega, Georgia, site of the first American gold rush. Second, it is a combination of a commercial gold mine and recreational gold mine. Visitors may take advantage of the gold-bearing earth recovered by their operation. Guests can pan for gold or search for gems. If you have any trouble ask the mine's grubbin' expert, a pet potbellied pig "Babe." There is a store on site, which features handmade 14 karat jewelry. A bathhouse, full RV hookups and primitive tent camping are available. If that is not enough there is a man-made lake stocked with bass, catfish and blue gills, which you can catch for $1.50 per pound. Oh yes, and there is a miniature golf course as well.

ADDRESS:
Clarence Norman, Jr.
Norman CE Wrecking Co.
3333 C E Norman Road
Lincolnton, GA 30817-3110
(706) 359-3862

DIRECTIONS:
From I-285, take 1-20 east to the Washington, GA exit SR78 (Route 10 and 17) Turn right onto SR 378 and drive 11 miles to Graves Mountain.

SEASON:
Year round by appointment.

HOURS:
Daylight hours only - must make prior arrangement with caretaker.

COST:
Donation to the caretaker.

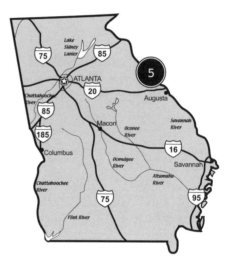

WHAT TO BRING:
Hand tools and containers for your finds. Ladders and power tools are not allowed.

INFORMATION:
This is a primitive site that has no facilities. You must bring your own food, drinks and supplies. This is best for experienced miners and rockhounds, children under 12 years of age are not allowed on the site. Call the caretaker to make an appointment to collect.

A wide variety of minerals can be found at this location by sifting through tailing piles or digging in likely locations. Among the various specimens that may be found here are: rutile, kyanite, lazulite, barite, sulfur, blue quart and quartz crystals with a hematite coating.

Please respect the code of conduct for this site by parking only in designated areas, stay away from high walls, no repelling off anything, and leaving the mountain by dusk.

ADDRESS:
Route 8, Box 540
Hidden Valley Road
Dahlonega, GA 30533

DIRECTIONS:
From Dahlonega, take U.S. Highway 52 East for two miles and turn left onto Rock House Road. Travel 1.6 miles and turn right onto Hidden Valley Road. Proceed 0.2 miles to the campsite.

SEASON:
March to December

HOURS:
Daylight

COST:
Free – Fee for panning and sluicing, if camping
$2 per day – Fee for panning and sluicing, if not camping
$5 per day – Fee for dredging, if camping
$10 per day – Fee for dredging, if not camping

WHAT TO BRING:
Some supplies are available for rent. Bring standard gold mining equipment. Dredges are permitted.

INFORMATION:
The privately owned Hidden Valley Campground is located in the peaceful north Georgia hills in an area that was once Cherokee Indian territory. In the 1500s, Spanish explorers traveled here in search for gold. They never found it, although it's here. Gold was first discovered in 1828 and is still being found today.

You can pan for gold, sluice or dredge in the two gold-bearing streams. Keep an eye out for gemstones such as garnets while you pan. The campground has twenty-two spaces for camping and RVs. Tent camping charges are $8 per night and RV sites with electric are $12 per night. Weekly and monthly rates are available. Facilities include restrooms, showers, and nature trails. Campers may search for gold for no additional charge.

MUSEUMS

7.
Dahlonega Courthouse Gold Museum
State Historic Site
Public Square
Dahlonega, GA 30533
(706) 864-2257
http://www.dahlonega.org/aboutus.asp
1828 was the year gold was discovered in Cherokee Territory. The courthouse serves as a historical museum. Look closely at the building's bricks. They contain trace amounts of gold trapped in the mud used to form them. The museum includes a short film describing life and mining techniques of the day. Nuggets, gold bearing quartz and coins minted here are on display.

8.
Georgia Museum of Natural History
University of Georgia
Athens, GA 30602-1882
(706) 542-1663
http://museum.nhm.uga.edu/
The geology collection contains over 20,000 specimens including 1,500 rocks & minerals from around the world. The paleontology collection exhibits over 12,000 fossils and casts.

9.
William Weinman Mineral Museum
51 Mineral Museum Drive
White, GA 30184
(770) 386-0576
http://www.weinmanmuseum.org
The Georgia Room houses rocks and minerals indigenous to this state. The Weinman Room exhibits specimens from around the world. There is also a fossil room and a discovery room especially designed to teach children about rocks and minerals. The museum also has a simulated cave with waterfall.

Impressions from the past

Illinois has coal and plenty of it. This state mines coal and quarries sand, gravel, portland, crushed stone and absorbent clay. Absorbent clay? That's right. The clay is used to remove contaminants. The fine particles soak up oil and water from garage and gas station floors. In the 1950s a new use was found for this product when the clay was substituted for sand in cat boxes. Kitty litter was born and is still going strong. Next time you have to clean that box, you can thank Illinois and that absorbing clay.

Below the clay, Illinois has shale and coal. Shale beds sit above and below the coal seams as evidence of the inland sea's advance. The shale was once the earth in which plants and forests grew. The coal once was the forest and swamp – vegetative matter is the makings of coal, hardened by time and pressure. Coal mining began in the 1860s in Illinois to meet local demands. By the end of the 1860s boomtowns and new railroads brought this coal to Chicago. Underground mining was and is dangerous work. On February 16, 1883, Braidwood, Illinois was the site of a disaster. A mine called the Diamond Mine was situated on marshy land, so when the snow above coupled with heavy rain seeped into the ground, the main passage flooded with alarming speed trapping seventy-four men below ground. Miners and family members above watched in horror as water bubbled up to ground level. Today there is a monument at the sight of this tragedy.

Some of the mines now are aboveground, keeping miners safer, but reducing their numbers. Mining this way involves stripping off layers of slate and discarding them. This process has proved a boon for paleontologists. The shale piles from active and abandoned area mines along Mazon Creek yields odd nodules of iron carbonate rock. Trapped within can be coal, calcite, siderite or fossils. These fossils are known worldwide by collectors because they preserve plants, insects, marine life and land animals, which have left no other trace. Sediment fell to the swamp bottom carrying with it leaves, animals and insects. Over time nodules, ranging from an inch to a foot long, formed around them. As the organisms rot away they left impressions, sometimes with intricate detail. Impressions show the complete creature, not just the skeletal system. The imprint is the weakest part of the nodule, so a well-placed hammer cleaves the rock in two exposing the fossil within. Some of the finds include: shrimp, ferns, trees, spiders, fish, early pine needles and cones, shells and some organisms found nowhere else. Over 400 plant species have been recovered here. For a look at some of these treasures visit the Field Museum of Natural History or the Illinois State Museum.

Fossils are not all that is worth collecting in Illinois. This state produces more fluorite than any other. Because of this, fluorite was made Illinois' state mineral. Found in colors of purple, blue, green, yellow and the pure form – clear, this mineral has many uses. Native Americans found fluorite in southeastern Illinois and used it for carving. Modern residents use this same crystal in the smelting process of ores, in manufacturing glass and enamel glazes, to help refine aluminum, even in the making of rocket fuel. Most people know that fluorite is found in toothpaste and used as a dental treatment, but don't know this same substance is used in concrete, cleaning solvents, anesthetics and degreasing agents.

Rockhounds visiting Illinois will enjoy searching for geodes, quartz and calcite in Hancock County and agates on the shores of Lake Superior.

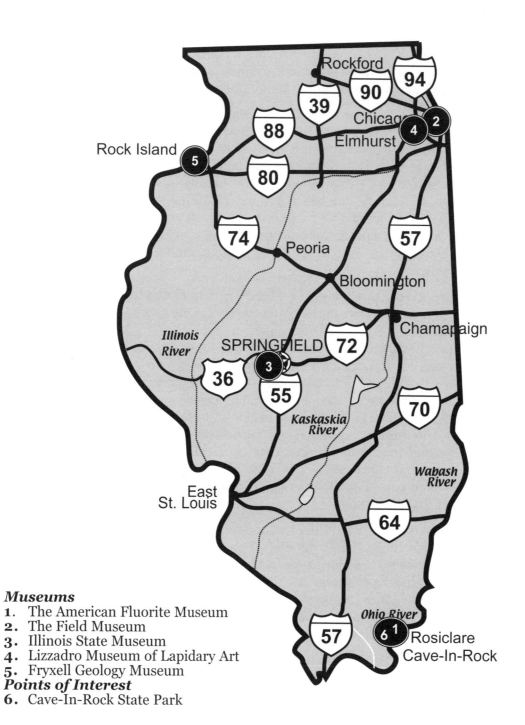

Museums
1. The American Fluorite Museum
2. The Field Museum
3. Illinois State Museum
4. Lizzadro Museum of Lapidary Art
5. Fryxell Geology Museum

Points of Interest
6. Cave-In-Rock State Park

MUSEUMS

Site 1.
The American Fluorite Museum
Main Street
P.O. Box 755
Rosiclare, IL 62982
(618) 285-3513
This small museum exhibits an impressive fluorite collection in a range of spectacular colors, some over a foot long. The museum also has items representing the fluor spar mining industry, mining paraphernalia and photographs.

Site 2.
The Field Museum
1400 South Lake Shore Drive
Chicago, IL 60605-2496
(312) 922-9410
http://www.fmnh.org/
Visit the permanent exhibits of minerals and rocks including the Grainger Hall of Gems. Most noteworthy at the Field Museum is Sue, the world's largest and most complete T-Rex.

Site 3.
Illinois State Museum
Spring & Edwards Streets
Springfield, IL 62706-5000
(217) 782-7387
http://www.museum.state.il.us/geninfo/index.html
The geology collection has over 200,000 specimens, including fossils and 6,700 rocks and minerals.

Site 4.
Lizzadro Museum of Lapidary Art
220 Cottage Hill
Elmhurst, IL 60126
(630) 833-1616
http://www.elmhurst.org/culture/lizzmus.asp
Lapidary, or cutting of stones, is the focus of this museum. Lizzadro claims to have the largest collection of Chinese jade carvings in the country. Exhibits include animal dioramas, mineral specimens, world famous diamonds and gemstones, and push button exhibits of geology.

Site 5.
Fryxell Geology Museum
Augustana College
Department of Geology
639 38th Street
Rock Island, IL 61201
(309) 794-731
http://helios.augustana.edu/academ/geology/fryxell.htm
The Fryxell Geology Museum at Augustana College is the second largest geology museum in the state and includes specimens of rocks and minerals from around the world, a fine collection of Illinois minerals, plus several interesting fossils.

Points of Interest

Site 6.
Cave-In-Rock State Park
Box 338
New State Park Road
Cave-In-Rock, IL 62919
(618) 289-4325
http://dnr.state.il.us/Lands/Landmgt/PARKS/R5/CAVEROCK.HTM
This historic cave was carved from the limestone by water and was used through the 1700s as a lair for thieves who preyed on travelers on the river below. In the 1830s pioneers used the well-known spot as a resting place on their long journey west. Today the State of Illinois maintains this cave for visitors to explore and be thankful they did not arrive three hundred years earlier.

Ancient seas, prehistoric swamps and mountains of ice.

Where did all that coal, limestone, gypsum and clay come from?

There are plenty of all these minerals in Indiana. Her quarries produce more dimension stone and masonry cement than any other state in the union. Other important minerals found here include gypsum used in creating Sheetrock, lime and clay. Finally coal is a valuable natural resource. Indiana has five surface and sixteen underground coal mines producing over 40 million tons annually.

The clay is easy to explain. Tremendous amounts of mud were left behind after the last Ice Age. Enormous glaciers tore into the limestone, gouging out rock and leaving clay as they receded.

The limestone formed in the Mississippian Period over 300 million years ago when Indiana was an ancient sea. Now known for its strength, color and durability, this limestone was once a collection of seashells and coral. Remains of billions of marine animals collected on the ocean bottom in vast quantities. Fossils were buried as well. One of the most common is the Crinoid, the state fossil. This creature resembled a plant and is closely related to starfish. The remains of the stem or spine of a primitive animal are often found in this state and are identified by the small donut shape, like a button-sized bead. Indiana supplies museums around the world with Crinoids.

Quarries now harvest ancient coral reef making Indiana the world's leading producer of limestone. The Empire State Building in New York City used Indiana's limestone in construction. To see this ancient seabed where it formed try McCormick's Creek State Park. Indiana's first state park has a crack in the limestone, called Wolf Cave, leading from one side of a hill to the other. Another place to enjoy ancient coral reefs is the Falls of Ohio. Here visitors can explore exhibits explaining the landscape and its origins before wandering past odd formations in limestone.

The coal and peat found in this state are also a remnant of another time and place. Shallow swamps covered much of this land and these ecosystems dropped large amounts of vegetative debris, which collected but could not rot, due to a lack of oxygen in the water. The plants were covered and pressed into peat and coal over millions of years. A natural gas, called methane, is produced in this process. Often it escapes but occasionally the gas collects in pockets above the coal. This is unfortunately true for much of Indiana's coal. Methane is colorless, odorless and tasteless. This gas can kill a miner underground, without warning. It is also very combustible which makes it the cause of some of the worst explosions and mine fires in this state. Just a rock falling on metal equipment may cause the spark that ignites this deadly unseen vapor.

Indiana proves that not all coal is black. Some of the coal here is brown or gray. The most unusual is an iridescent specimen of blue and red called peacock coal. The colors are created by a thin film of iron oxide on the coal, much like oil on a mud puddle. Regrettably the effect is temporary and disappears shortly after exposure to the surface.

So when you gaze out on the cornfields of Indiana, imagines the inland sea, ancient swamps and massive glaciers that were here long before.

Diggings:
1. Yogi Bear's Jellystone Park
2. Interstate 64 Roadcut

Museums
3. Indiana State Museum

Caves
4. Bluesprings Caverns
5. Marengo Cave National Landmark
6. Spring Mill State Park
7. Squire Boone Caverns
8. Wyandotte Caves

Points of Interest
9. Angel Mounds State Historic Site
10. Falls of the Ohio State Park
11. McCormick's Creek
12. Mounds State Park

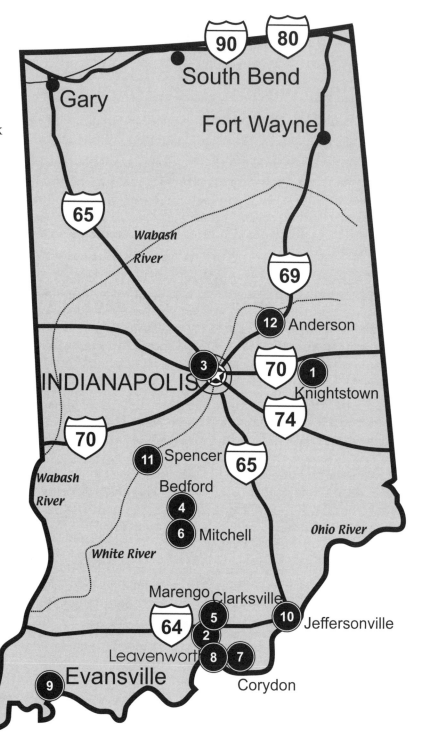

ADDRESS:
Jellystone Park at Knightstown
5964 South Street Road 109
Knightstown, IN 46148
(765) 737-6585 or (800) I-GO-YOGI
info@jellystoneindy.com
http://www.jellystoneindy.com/goldcamp.html

DIRECTIONS:
Located on State Highway 109 one-half mile
north of Interstate Highway 70 from Exit 115.

SEASON:
Open all year round – digging as weather
permits.

HOURS:
Call for information

COST:
If camping, the prospecting fee is $3.00 for
adults and $2.00 for children. If you are not
camping the fee is $6.00 and $2.00 respectively.
Camping costs between $21.00 and $30.00.
Cabin rentals are up to $65.00.

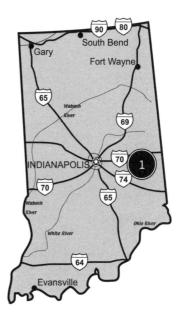

WHAT TO BRING:
Bring your own prospecting equipment or rent equipment from the park.

INFORMATION:
   This is a recreation of a 1860s gold camp. This site is a Gold Prospectors Association of
America (GPAA) registered claim. Free panning lessons are available and equipment available
for rent. Those who have their own equipment may use it. Prospectors may use high-bankers,
sluice boxes, and dredges up to four-inches. Digging in the banks is not permitted. Dig all you
want in the creek bed or gravel beds.
   Camping sites are available for all types of camping. Some cabins are for rent. There is a
camp store, pool and paddleboat rentals.

ADDRESS:
Interstate Highway 64
Exit 86
Crawford County, IN
No phone

DIRECTIONS:
Take Interstate Highway 64 to Sulphur, Indiana in Crawford County, and get off at Exit 86. The road cut is directly off the exit.

SEASON:
Open all year round

HOURS:
Daylight

COST:
Free

WHAT TO BRING:
Bring standard mining tools, your own food and plenty to drink. Also include a collecting bucket, small rock hammer, prying tools, hand garden tools, and a small broom or brush.

INFORMATION:
　　This site is a bit different. It is not a private mine or public park, it is a roadcut. Unearthed by the construction of Interstate Highway 64, this area is rich in fossils. It has become a very popular site for local rock hunters. Since it is located along the exits of Interstate Highway 64, there is no organization or phone number listed above to provide additional information. But if your travels take you along Interstate Highway 64 you may want to take Exit 86 to have a look around.

　　This area contains many types of fossils including crinoids, fish remains, horn coral, shark's teeth, snails and brachiopods. Material is found in shale and limestone deposits. The best areas to find fossils are on the northwest and northeast corners of the exit. There is plenty of parking. Please respect the area and other collectors. Collecting is for personal use only. Search at your own risk.

MUSEUMS

Site 3.
Indiana State Museum
650 West Washington Street
Indianapolis, IN 46204
(317) 232-1637
http://www.in.gov/ism/index.asp
The minerals in this museum range in size from less than an ounce to several hundred pounds. Over 2,000 specimens displayed included most minerals found in Indiana.

CAVES

Site 4.
Bluespring Caverns
1459 Bluespring Caverns Road
Bedford, IN 47421
(812) 279-9471
http://www.bluespringcaverns.com/AreaMap.html
In the 1940s a farmer's pond disappeared overnight and the entrance to Bluespring Cavern was discovered. The White River now flows only a short walk from the cavern entrance where a boat tour of this unique underground tour unfolds. Blind fish and crayfish thrive in the underground environment. A guided boat tour takes guests through a unique subterranean world of limestone and water, revealing the awesome forces of nature and time.

Site 5.
Marengo Cave National Landmark
Box 217 Marengo Cave Road
Marengo, IN 47140
(812) 365-2705
http://www.marengocave.com/
Underground geologic riches await visitors to the Marengo Cave. Visitors on the Crystal Palace tour travel through huge corridors of massive flowstone formations. Looking for more adventure? Try the two-hour Old Town Spring Cave tour for those who want to learn safe caving techniques and get muddy.

Site 6.
Spring Mill State Park
Box 376
Mitchell, IN 47446
(812) 849-4129
http://www.in.gov/dnr/parklake/parks/springmill.html
Twin Caves can be explored by boat or by foot. A restored pioneer village, fishing, picnicking, hiking and camping are available within the park.

Site 7.
Squire Boone Caverns
PO Box 411
Corydon, IN 47112
(812) 732-4382
http://www.squireboonecaverns.com
Named for a prominent pioneer, explorer and Revolutionary War veteran, Squire Boone, this cavern offers tours every thirty minutes. Squire Boone and his brother Daniel discovered this cave in 1790. He eventually settled his family here and after his death he was buried within the cave he loved, as was his request. Guided tours take visitors past unique cave formations and underground waterfalls. The gift shop has a very large rock shop, which imports two hundred tons of mineral annually. Visitors may also enjoy sluicing for gems and gold on site.

Site 8.
Wyandotte Caves
7315 S. Wyandotte Cave Road
Leavenworth, IN 47137
(812) 738-2705
http://www.wyandottecaves.com
This cave has the distinction of offering visitors six different tours from which to choose and includes views of famous limestone formations. Tours range in length from thirty minutes to seven hours. Serious spelunkers will enjoy the challenge of long crawls and some climbing.

POINTS OF INTEREST

Site 9.
Angel Mounds State Historic Site
8215 Pollack Avenue
Evansville, IN 47715
(812) 853-3956
http://www.angelmounds.org/main.asp
This is of one of the best preserved prehistoric Native American sites in the country. The location was a stockade around a town inhabited from 1100 to 1450 A.D. The actual mound is 600 feet long, 400 feet wide and forty-four feet high making it a formidable structure. Those interested in archeology and history will find plenty to consider at Angel Mounds.

Site 10.
Falls of the Ohio State Park
201 West Riverside Drive
Clarksville, IN 47129
(812) 280-9970
http://www.fallsoftheohio.org/
This state park sits on a 350 million year old coral reef left behind by an ancient inland sea. In honor of its past, this park includes an large saltwater aquarium complete with coral reef. Other activities include hiking along the exposed coral reef, boating, educational programs and visiting the interpretive center.

Site 11.
McCormick's Creek
Route 5, Box 282
Spencer, IN 47460
(812) 829-2235
http://www.state.in.us/dnr/parklake/parks/mccormickscreek.html
The unusual limestone formations and lovely waterfalls bring visitors back again and again to Indiana's first state park. Hiking trails take guests past canyons and beneath high cliffs. Visit the nature center or join an interpretive hike. Camping, hiking, swimming and tennis are popular activities here.

Site 12.
Mounds State Park
4306 Mounds Road
Anderson, IN 46017
(765) 642-6627
http://www.state.in.us/dnr/parklake/parks/mounds.html
This park contains ten earthwork structures built in prehistory by Indian tribes. Archaeologists believe this site was a gathering place for religious ceremonies. The park also contains a nature center, camping facilities, hiking trails, pool and picnicking sites.

Lead into bullets for the Civil War

The limestone and shale found below the soil lies in silent testament to this region's previous existence as ocean bottom. The limestone is built from coral reefs and marine mollusks that left their shell and skeleton behind when the shallow sea retreated and the prairie appeared. This limestone now is quarried. The fossils found in Iowa's road cuts, stream banks and other exposed areas bear witness to the slow march of time. Coral, shells from snail and mollusk lie with primitive fish and sponges turned to stone. Crinoid fossils are found in LeGrand and Burlington. Seeds and ferns appear in fossil form. Above them, Mammoth teeth show a time when the ocean disappeared and the world turned to ice.

The glaciers did not reach Iowa, but debris reached this state. Tons of material dragged from the volcanic area of Lake Superior moved with the water from the melting glaciers and came to rest in Iowa. Rocks that traveled far from their origins include geodes, formed in lava bubbles and filled with sparkling quartz, agates, chalcedony, opal and petrified wood.

Much of Iowa is prairie grasslands, devoid of trees. Many of the trees that did grow in the eastern and northern portions of the state were rapidly timbered for the building of homes. Lumber was so sparse that settlers built houses from sod, cut from the prairie and used stone or dirt to create fence posts. Hay, corncob and animal droppings became fuel. One can see why the discovery of local coal was such welcome news.

Soon settlers discovered another resource. Local tribes mined lead for ornaments and functional objects as evidenced by the galena found in burial mounds throughout the state. Europeans first mined here in 1650 when French fur traders recovered and smelted the ore for bullets. By the 1780s activities were organized so the lead was mined, smelted and then shipped to market. Iowa was a major source of lead for bullets in the Civil War.

Local tribes continued to mine into the eighteenth century in pits and caves. Indians who could not hunt in the summer, mined and traded with the settlers gaining valuable goods for their tribes. A French-Canadian named Julien Dubuque received use of a lead mine from the Meskwakis tribe in 1788. Dubuque called this site "the Mines of Spain." This site is now a state park of the same name. Over the next two decades, he made a killing mining ore and shipping the lead to St Louis. Dubuque died in 1820, possibly from lead-poisoning. Land speculators rushed to find the site of his mine, but the Meskwakis tribe resumed ownership of their loaned land and defended the tract from interlopers. Appropriately, Dubuque was buried on a bluff close to his mine in a wooden tomb along the river. For many years the place was a landmark and local attraction on the Mississippi.

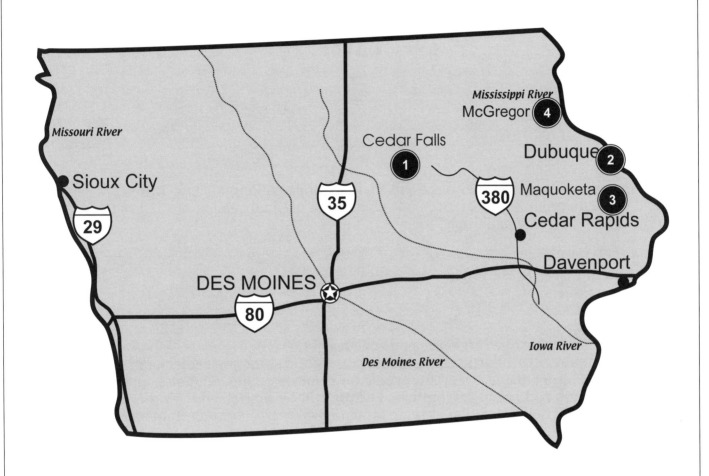

Museums
1. University of Northern Iowa Museum
Caves
2. Crystal Lake Cave
3. Maquoketa Caves State Park
4. Spook Cave

Museums

Site 1.
University of Northern Iowa Museum
3219 Hudson Road
Cedar Falls, IA 50614-0199
(319) 273-2188
http://www.uni.edu/museum/
The University Museum is a natural history museum and includes geology exhibits.

Caves

Site 2.
Crystal Lake Cave
7699 Crystal Lake Drive
Dubuque, IA 52003-9504
(563) 556-6451
http://www.crystallakecave.com
Lead miners made an unusual discovery in 1868 while sinking a shaft. Instead of ore, they found an underground geologic wonder. One of the original miners opened the cave to the public in 1932 and it remains open today.

Site 3.
Maquoketa Caves State Park
10970 98th Street
Maquoketa, IA 52060
(563) 652-5833
http://www.state.ia.us/dnr/organiza/ppd/maqucav.htm
This cave contains ancient artifacts including pottery and stone projectile points revealing that this place has been a popular stop for people for countless years. Settlers happened upon the site in the 1830s and began exploration. Visitors will be impressed with the stalactites that remain in spite of the destruction of some formations by early souvenir hunters.

Site 4.
Spook Cave
13299 Spook Cave Road
McGregor, IA 52157
(563) 873-2144
http://www.traveliowa.com/media_center/details_column.htm?seq=11
The sounds coming from the bluff were not from spooks, but water traveling through the cave. Visitors explore by a guided boat tour as guides explain the caves formation, discovery and early history.

Not for all the gold in Fort Knox

Kentucky relies on mining as one of the foundations of the state's economy. Surface mining operations include clay, limestone, sand and gravel. Other mineral resources found here include: fluorite, barite, lead, calcite and sphalerite. Coal mining first began in the mid-1700s and reached one million tons per year just before the 1900s. Kentucky led the nation in coal production and was only recently displaced by Wyoming in 1988.

Pearls are as far from coal as one can imagine, but Kentucky has both these resources in common. Freshwater pearls are found here and known for the wide range of colors including: white, silvery-white, pink, red, bronze, lavender, green, blue and yellow. Colors are a result of the mussel species, position of the pearl in the shell and the water quality. Technology for culturing freshwater pearls became available in the late 1970s using mussel shell from the southeastern U.S. as seeds. Supplying raw material for the creation of cultured pearls is a thriving, if dangerous, business in Kentucky's rivers and lakes. But these same bodies of water are used for other things as well.

Kentucky third largest revenue producer is tourism, with good reason. This state has six national sites and almost 50 state parks. One of the most famous is Mammoth Cave. This is the longest known cave in existence and was opened in 1816 for tourism, making it the second oldest attraction in the United States, behind Niagara Falls. This amazing natural feature is now a National Park and worth a visit.

Kentucky also has gold – tons and tons of gold. Unfortunately it is stored in a classified facility and not a mining camp. Fort Knox holds over 140 million ounces of gold for the United States government, but they do not permit visitors. The actual building was constructed in 1936 of granite, concrete and reinforced steel and is guarded by United States treasury agents. The site quickly became synonymous with impregnability, as in "locked up a tight as Fort Knox." As a secure site, Fort Knox has at times protected some other national treasures including the United States Constitution, the Magna Carta, Lincoln's Gettysburg Address and the Declaration of Independence. But unfortunately they don't give tours and they don't provide free samples. Sorry – no exceptions!

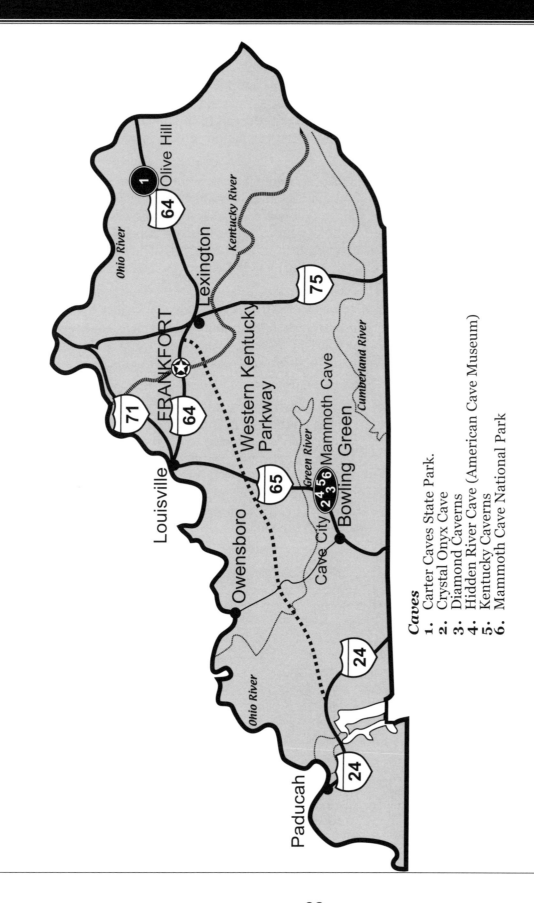

Caves
1. Carter Caves State Park.
2. Crystal Onyx Cave
3. Diamond Caverns
4. Hidden River Cave (American Cave Museum)
5. Kentucky Caverns
6. Mammoth Cave National Park

CAVES

Site 1.
Carter Caves State Park
Carter Caves State Resort Park
344 Caveland Drive
Olive Hill, KY 41164-9032
(606) 286-4411
http://www.state.ky.us/agencies/parks/cartcave.htm
This lovely state park contains twenty caverns, three of which are open for tours year round. Cascade Cave is noted for its beauty and a thirty-foot waterfall. X Cave gained its name from the crossing pattern of the cavern's passages and Saltpetre Cave was used by settlers as a mine. The largest cave is Bat Cave, winter home of thousand of bats and so is open to human visitors only in the summer months.

Site 2.
Crystal Onyx Cave
8709 Happy Valley Road
Cave City, KY 42127
(270) 773-2359
http://www.crystalonyxcave.com
Pristine formations greet visitors in this cave opened in the 1960s. Native Americans used this place over two thousand years ago as a burial site. The cave has rare species including blind crayfish, adapted to total darkness. Hour long tours depart regularly throughout the day.

Site 3.
Diamond Caverns
1900 Mammoth Cave Parkway
Park City, KY 42160
(270) 749-2233
http://www.diamondcaverns.com
South Central Kentucky, adjacent to Mammoth Cave National Park, is the location of this cave open to the public for over one hundred and forty years. Tours travel about a half-mile and take about one hour past drapery deposits, flowstone, colorful calcite and thousand of stalactites and stalagmites.

Site 4.
Hidden River Cave (American Cave Museum)
P.O. Box 409
119 East Main Street
Horse Cave, KY 42749
(270) 786-1466
http://www.cavern.org/
The huge entrance to this cave begins seven miles of passages connected by an underground river dubbed "Hidden River." Begin your trip at the American Cave Museum for information on history and science before beginning your guided tour.

Site 5.
Kentucky Caverns
Kentucky Down Under
Kentucky Down Under/Kentucky Caverns
3700 L. & N. Turnpike Road
P.O. Box 189
Horse Cave, KY 42749-0189
(800) 762-2869 or (270) 786-2634
http://www.kycaverns.com
Known for the unique onyx formations, this cave tour features stalagmites, stalactites and flowstone. There are some steps on this guided tour.

Site 6.
Mammoth Cave National Park
P.O. Box 7
Mammoth Cave, KY 42259
(502) 758-2328 or (270) 758-2180
http://www.nps.gov/maca/
Mammoth Cave is aptly named and known to be one of the longest surveyed caves in the world. Over 365 miles of passages have been explored to date and the discovery continues. This sprawling behemoth burrows beneath five different ridges and exists at five different depths. The National Park allows visitors access to over ten miles of passages. Tours average one to two miles and are moderately strenuous.

Prospectors used airplane propellers to drive away mosquitoes

The earliest human inhabitants of this area hunted mastodon over 10,000 years ago as evidenced by stone tools found beside skeletal remains of the massive beasts. These hunters were, no doubt, the first to slap at mosquitoes as well. This state emerged from the sea too late to collect dinosaur remains, but there are fossils from the remains of thirteen million-year-old creatures including a six-foot tall beaver, elephants, camels, rhinos and smaller animals. Another remnant of an earlier time is the petrified wood found throughout Louisiana, Texas and Arizona.

This state has an abundance of water in the form of slow moving rivers and streams that trickle through bayous, which flow from the Mississippi to the Gulf of Mexico. The City of New Orleans is partially below sea level and uses pumps, walls and levees to keep dry. Sediment from the rivers is deposited along banks and coastal plains thus increasing the size of the state.

The official symbols of Louisiana reflect the presence of water here. The state tree, the bald cypress loves swampy coastal marshes and the state bird is the pelican, of course.

Louisiana is best know for products like rice and seafood. Both fishing and agriculture are vital to the state's economy. But mining is important as well.

Sulphur, used to create sulphuric acid for use in petroleum refining, chemical production, steel production, paint insecticides and other common products is found here in abundance. Unfortunately the sulphur lies beneath swamps and muck. Many men died trying to sink shafts through this unstable ground. In the 1890's a way to extract valuable sulphur was discovered. Miners sent superheated water into the deposits where the liquid dissolved the sulphur so it could be pumped to the surface. Louisiana now produces over two million tons of sulphur annually.

Working in swamps and bayous is difficult. Swarms of mosquitoes make workers miserable. So in 1930 airplane propellers powered by Model-T motors worked as blowers keeping the biting hordes at bay. Offshore drilling began in the mid-1950s and continues today. Moving out to sea is one way to escape the biting insects!

Louisiana is the third largest producer of petroleum and also drills for natural gas. Another important resource is salt. It does not come from the Gulf of Mexico, but rather from a much earlier sea. The huge underground salt deposits are 50,000 feet deep in places and produces pure rock salt.

Louisiana has petroleum, natural gas, rock salt and sulphur, making use of resources hidden beneath land, bayou and gulf waters.

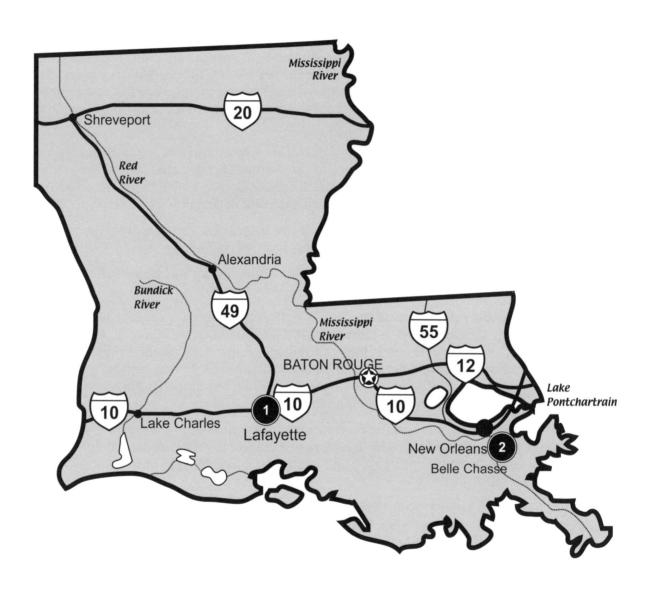

Museums
1. Lafayette Natural History Museum
2. Tulane University Museum of Natural History

Museums

Site 1.
Lafayette Natural History Museum
433 Jefferson Street
Lafayette, LA 70501
(337) 291-5544
http://www.lnhm.org/
This museum includes cultural artifacts from Louisiana's past with many Native American items, fine arts, decorative arts, fossils including a mastodon and meteorites.

Site 2.
Tulane University Museum of Natural History
Bldg. A-3, Wild Boar Road
Belle Chasse, LA 70037
(504) 394-1711
http://www.museum.tulane.edu/
This museum is open for arranged school tours only and includes a collection of invertebrates, fishes, amphibians, birds, mammals and reptiles. There is also a collection of vertebrate fossils.

The first gemstone mined in North America

People know about Maine tourmaline. These rare gemstones are recognized worldwide for unique beauty. No other gemstone occurs in such a wide range of colors. Credit for discovery goes to two boys, named Elijah Hamlin and Ezekiel Holmes, who noticed several green stones trapped in the roots of an overturned tree in the year 1820 in the area of Paris, Maine. Mines sprang up and many are still going strong.

Tourmaline describes the crystal formation rather than a composition of the gem. This explains why there is such a range of colors in this category. Crystals can be black, white, green, red, blue or yellow. Black is the most common and least valuable form of tourmaline. The green gains color from iron, while the gemmy red comes from a concentration of manganese. These crystals, range in size from microscopic to several feet in length, and form in a type of granite called pegmatite. On rare occurrences the crystals form in pockets within the pegmatite. Gems growing in such pockets tend to be clear and of the finest quality. Perhaps most fascinating of all the tourmaline varieties are stones containing more than one color in the same crystal. Bi-colored crystals are often pink at one end and green at the other.

In 1972 large pockets of tourmaline were found in Newry, Maine. Hundreds of pounds of crystals were unearthed including what is now called watermelon tourmaline. This type of gem has an inner pink core surrounded by an outer green "rind" making a cross-section of the crystal look like a transparent slice of watermelon. Included in the 1972 glory hole was the 10-inch watermelon tourmaline crystal dubbed "the Jolly Green Giant." This astonishing crystal now resides at the Smithsonian Institution in Washington D.C.

Tourmaline forms in crystallized prisms with flat or wedge shaped terminations. The outer casing of these gems is hexagonal, making them resemble an unsharpened pencil. Each crystal forms with a different structure at each end and occasionally a different color. When heated the crystal becomes magnetic and is known to attract dust.

Several tourmaline mines are open to the public. Check the digging section to locate a likely mine. Those interested in obtaining a gem without digging may want to visit Mount Mica Rarities in Greenwood, Maine. There are many museums including the Maine State Museum, Nylander Museum and Perham's of West Paris where visitors can see specimen pieces of rough and cut tourmaline.

While tourmaline is the most famous of Maine's gemstones, it is by no means the only thing worth collecting here. Beryl, amazonite in green feldspar and topaz are all occasionally found along with several types of fossils. Most of the fossils come from the Mesozoic period and include marine shellfish such as clams, coral, brachiopods as well as snails and trilobites.

Shortly after tourmaline was discovered, Maine's first state geologist conducted a mineral survey discovering granite, limestone, slate and feldspar near the coast. Mineral resources near water made shipping easier. Quarries were opened, rock cut and sent up and down the east coast. When iron ore was discovered, the Newfield Iron Works began mining and smelting the metal.

After the California Gold Rush, many local miners returned to Maine. They discovered outcroppings in their home state similar to those seen in the gold fields. Silver and lead were discovered and mined, but the veins were not very rich and harsh conditions in the Northeast including the winter snows caused great hardship. In the spring the snowmelt flowed into the mines hindering progress. There was a silver rush from 1878 to 1882 when at least twelve mines operated. Most mines could not turn a profit and closed.

During the 1940s manganese, on the War Departments strategic list, was actively mined in Maine. Nickel and copper deposits were discovered in the 1960s and mined, along with zinc until the mid-1970s.

Gold is found over a wide area in Maine. Prospectors generally search in the gravel of rivers and streams. Some nuggets are found here. Panning, sluicing and use of a dredge up to four-inches does not require a permit in Maine. Be sure to ask permission of any private property owners and check park and forest regulations before you mine.

The rocky Maine coast near Penobscot Bay and Eastport yield yet another source of treasure. Agates and jaspers, smoothed by tumbling in the ocean, require nothing more than scooping them up and placing them in a pocket. Try Jasper Beach for a start.

With all the variety of rocks, gems and minerals available in Maine, the only question is where to begin?

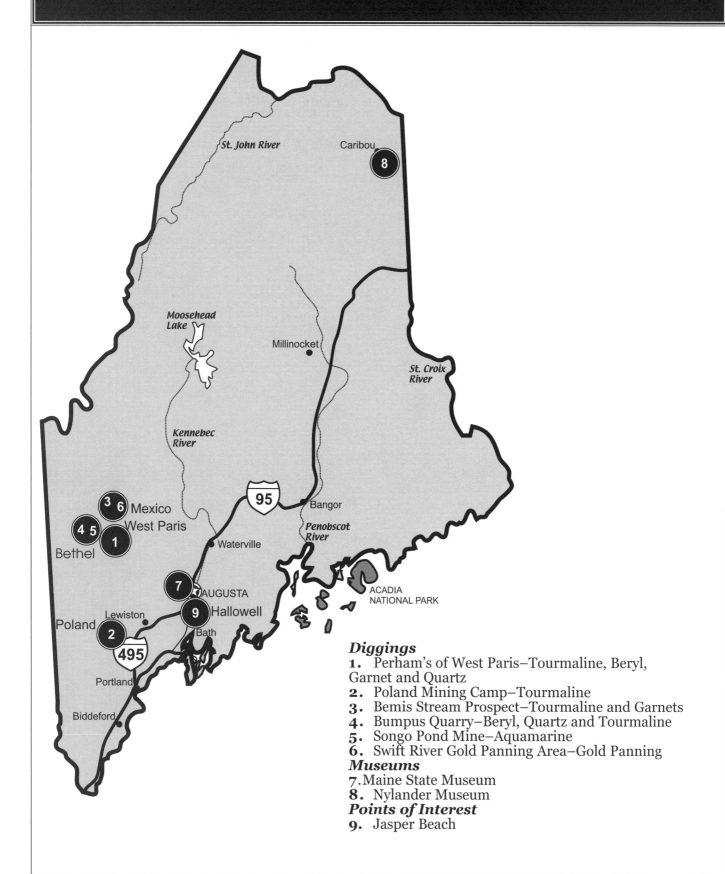

Diggings

1. Perham's of West Paris–Tourmaline, Beryl, Garnet and Quartz
2. Poland Mining Camp–Tourmaline
3. Bemis Stream Prospect–Tourmaline and Garnets
4. Bumpus Quarry–Beryl, Quartz and Tourmaline
5. Songo Pond Mine–Aquamarine
6. Swift River Gold Panning Area–Gold Panning

Museums

7. Maine State Museum
8. Nylander Museum

Points of Interest

9. Jasper Beach

ADDRESS:
Perham's of West Paris
P.O. Box 280
West Paris, ME 04289
(800) 371-GEMS or (800) 371-4367 or (207) 674-2342
http://homepage.mac.com/rasprague/PegShop/perham.html

DIRECTIONS:
The Perham's Store is located on Route 26 in West Paris.

SEASON:
Open mid-May to mid-October, weather permitting.

HOURS:
Store hours 9 a.m. to 5 p.m., seven days per week

COST:
Some quarries are free and others charge a nominal fee.

WHAT TO BRING:
Bring all your own equipment including: buckets, picks, rock hammer, gloves and don't forget bug spray.

INFORMATION:
The store/museum maintains several collecting quarries in the area. Maps and information are available at the store. Take a few minutes to study the museum so you will know how to recognize various rocks and minerals. The store sells gemstones, jewelry, books and other gifts.

The collecting quarries are open pits where visitors look through tailing piles or dig new material depending on your energy level. Each quarry is known for different minerals so if you are searching for a specific rock, ask the owners of the store where best to look.

Most of these quarries were once mined for mica and feldspar. Other minerals are found in association with these in a variety that is truly amazing. Look for green tourmaline, orange and smoky quartz, garnet, beryl, black tourmaline, purple apatite, mica, spodumene, rose quartz and amethyst.

ADDRESS:
Poland Mining Camps
P.O. Box 26
Poland, Maine 04274
(207) 998-2350
http://users.rcn.com/kenx/poland.htm

DIRECTIONS:
Take Exit 11 from Interstate Highway 495, Gray, Maine. Go north on Route 26 through the town of Poland, Maine. The Poland Mine is on the right across from the Poland Regional High School. If you see Route 11 you missed it.

SEASON:
Memorial Day to September 30

HOURS:
Collecting trips departs from Poland Camp at 8:00 a.m.

COST:
Reservations required. Prices are based on the length of your stay and the type of accommodations you choose. Packages include three meals a day and lodging. Tent sites, RV and full service cabins are available.
Cabin rental/meals and mining -$472 per person per week.
Rates are lower for RV and Tent Campers.
$45 per person per day for day-tripper (miners not staying at Poland Camp)
$85 per person per two-day weekend for day-trippers.
Three-day minimum to lodge at Poland Mining Camp Cabins.

WHAT TO BRING:
All the tools you need to mine, hat, gloves, work boots, eye protection, bug spray, collecting buckets.

INFORMATION:
This trip includes lodging and all your meals, plus access to many working mines not open to the public including Mount Mica, Emmons Quarry and Mount Apatite. Day-trippers are welcome when space is available. Poland Camp welcomes beginners and advanced miners. Their operation is ideally located in the heart of Maine's mineral collecting area and pegmatite quarries. Most locations have freshly exposed material each week allowing miners great opportunities for finding minerals. Guides are provided with each excursion.

Full service cabins sleep up to four adults and include a living room with fireplace, kitchen, full bathroom and two bedrooms. RV and tent sites are available. A new pavilion serves as the gathering place for meals and get-togethers. Meal package includes a continental breakfast, a pack-your-own lunch and a buffet supper. Local craftsmen give lectures and demonstrations here in the evenings. Laundry facilities and a pay phone are available. Children and pets are welcome.

ADDRESS:
Mexico Field Office
9 Main Street
Mexico, ME 04257

DIRECTIONS:
From the town of Mexico, at the junction of U.S. Highway 2 and State Highway 17, drive north 17.1 miles on State Highway 17 to the town of Houghton. Turn left after the field, onto the dirt road and cross the bridge at Swift River. Go northwest on the gravel logging road for 6.3 miles from State Highway 17 to the bridge over Bemis Stream. Park on the right.

SEASON:
Summer and Fall

HOURS:
Daylight hours

COST:
Free

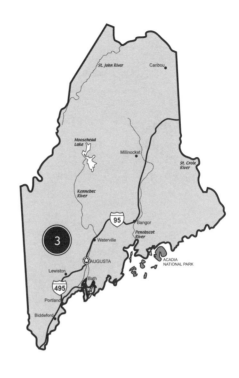

WHAT TO BRING:
Bring standard mining tools, your own food and plenty to drink. You may also need a quarter-inch screen and a container to hold your finds. Maine is noted for its beautiful woodlands and its huge mosquitoes and biting black flies. A high quality bug spray is recommended.

INFORMATION:
The Bemis Stream location is on federal land. You must call to get permission to prospect here. Only small scale prospecting and personal collecting is permitted. Permission is easily granted. Green tourmaline crystals can be found in the ledge along the river and garnets and other quartz can be found here as well.

ADDRESS:
Rodney Kimball
P.O. Box 121
Bethel, ME 04217
(207) 836-3945 or (207) 824-2144

STREET ADDRESS:
Route 2
West Bethel, ME 04217

DIRECTIONS:
The Bumpus Quarry is located in Albany, Maine. You must call Mr. Kimball at his shop to make reservations to visit the quarry. His shop is located on U.S. Highway 2 in West Bethel.

SEASON:
Spring to late Fall

HOURS:
Daylight

COST:
$25 per person – Adults (depending on number in party)
Free – Children ages 8 and under

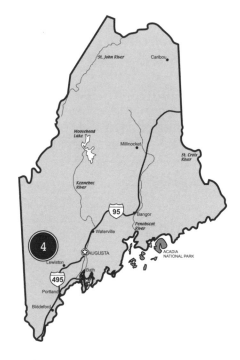

WHAT TO BRING:
Bring standard mining tools, your own food and plenty to drink.

INFORMATION:
The Bumpus Quarry is primarily composed of coarse-grained, light-colored granite. Millions of years ago, when the granite was gradually cooling, rare elements seeped into the fractures in the rock. These elements crystallized to form beryl, quartz, tourmaline, zircon and garnet deposits. This is a great place to pick up some nice specimens for your collection.

This location requires hard rock mining. You will need to break apart rocks found in the open pit dumps or work the quarry walls. Well-formed crystals grow in the granite, tightly encased in the rock matrix. The best crystals are found surrounded by quartz. Quartz is most commonly milky white, smoky gray or black. Look for quartz crystals and gently break away the brittle quartz to expose matrix specimens. Large flat sheets of mica in black and white are also common here.

Before heading out please call Mr. Kimball a few days prior to your trip to make reservations.

ADDRESS:
Jan Brownstein
P.O. Box 864
Bethel, ME 04217
(207) 824-3898

DIRECTIONS:
The mine is just south of Bethel, Maine. Take State Highway 5 and follow the signs to the Songo Mine. The mine is 4 miles from Bethel and 0.25 miles from the State Highway 5 turnoff.

SEASON:
May 1 to October 31 – Weather permitting

HOURS:
The mine opens at 9:00 a.m.

COST:
$10 – Adults ages 13 and older
$5 – Children (any "collecting" children)

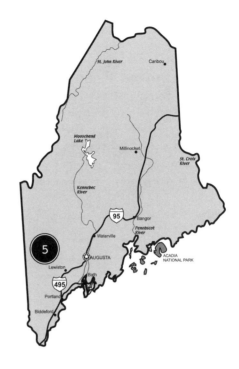

WHAT TO BRING:
Bring standard mining tools, your own food and plenty to drink.

INFORMATION:
This is a working mine that is currently in operation mining aquamarine. You must call ahead for an appointment to collect. The mine offers opportunities to watch the operation work. Collectors are permitted to work the tailings pile for aquamarine, apatite, tourmaline, mica, garnet and quartz. Some help is offered to beginners to get you started on the right foot. You may keep all you find, limited to one, five-gallon bucket of material per day.

The mine is located on a hill with a view of Songo Pond on one side and the Sunday River on the other. The tailings pile is usually in full sun, so a hat and sunscreen is recommended. You may drive tools and passengers to the mine, but then you must park in the lot.

ADDRESS:
Mexico Field Office
9 Main Street
Mexico, ME 04257

DIRECTIONS:
From Coos Canyon Picnic Area on State Highway 17 in Byron, drive 1.7 miles north and then east on Byron Road toward Tumbledown Mountain. Turn left onto the unpaved road. Take the fork to the right and cross the branch of the Swift River. Park your car on either side of the road after crossing the bridge.

SEASON:
Summer and Fall

HOURS:
Daylight

COST:
Free

WHAT TO BRING:
Gold Mining tools including a gold pan, sluice, dredge, bucket, shovel, pick, and work gloves.

INFORMATION:
    The Swift River is in Byron in Oxford County. You must get permission to prospect by calling or writing the Unit Supervisor at the above address. No state permit is required for small scale mining.
    Disturbance to the riverbank is prohibited. You may use a pan, sluice or a dredge with up to a four-inch hose diameter. Large scale mining requires a permit from the Maine Department of Environmental Protection.
    Flakes to nuggets have been found in this area, along with other heavy material including almandine garnet, magnetite, and staurolite. The area downstream of the parking area has produced gold.
    Maine is rivaled only by Florida and Alaska in the size and ferocity of its biting insects. Bring a strong, deep woods bug repellent. Long pants and long-sleeved clothing are suggested.
    The Timberlands manager reminds us that deer hunting season is in late November, bow season is earlier. Wear bright clothes or avoid mining during this time. If you do decide to collect during hunting season, keep children and pets close by.

Museums

Site 7.
Maine State Museum
83 State House Station
Augusta, ME 04333-0083
(207) 287-2301
http://www.mainestatemuseum.org
This museum includes a fine variety of minerals and gems from the state of Maine, including examples of the official state mineral, tourmaline.

Site 8.
Nylander Museum
657 Main Street
Caribou, ME 04736
(207) 493-4209
http://www.nylandermuseum.org/
This small museum includes minerals and fossils.

Points of Interest

Site 9.
Jasper Beach
Maine Publicity Bureau
P.O. Box 2300
Hallowell, ME 04347
(207) 623-0363
http://www.state.me.us/doc/nrimc/mgs/sites-2000/jun00.htm
This gravel beach contains plenty of jasper-like stone called rhyolite, a red volcanic stone commonly found here. Beachcombing is popular here and the rhyolite takes a nice polish.

Stone from Maryland graces two of Washington D.C.'s most famous structures

Maryland's stone and mineral wealth may surprise you. The state has a variety of stone suitable for quarrying. There is metal wealth as well in the form of copper, iron, gold and chrome. Some of the most accessible fossil riches on the east coast are found in Maryland. Let's begin with the mineral riches, which are not readily apparent.

Early settlers in western Maryland dubbed an area the serpentine barren. The greenish exposed stone supported little vegetation and no crops, so it was called "the barrens." Some of the serpentine was quarried for decorative building stone, but the real mineral wealth was overlooked by all but Isaac Tyson, Jr. who knew that chromite occurs with serpentine. In the early 1800s he noticed some green stones in a wagon, and set out to discover their origin finding the barrens. He also found what he sought – chrome ore. By 1845 Tyson owned a chrome plant near Baltimore, and Maryland led the world in chrome production. This bright metal is now used to strengthen steel and give metal a mirror shine. In the 1800s the metal was used to dye fabric, color pottery and glass, and in processing leather. In the late 1800s vast chrome deposits were discovered in Asia and production in Maryland ceased. Other metals were discovered.

Three valuable copper veins, mined in the 1700s, continued production, with a brief halt during the Revolutionary war. Maryland led the nation in copper production until the opening of the Lake Superior region.

The Civil War may be credited with the discovery of gold in Maryland. A private stationed outside Washington, D.C. found gold while washing a skillet. Several commercial mines opened unearthing a fine-grained gold, and occasionally a gold bearing quartz vein. As technology improved so did gold extraction. Mining continued through the 1940s. Perhaps as much as 5,000 ounces of gold was produced in this region. Gold mining is now the domain of hobbyists. If you plan to try your luck, be advised that much of this land is private and gold prospecting is prohibited on state owned land.

Maryland coal mining began in Colonial times accommodating local needs for fuel. Coal became commercially important in the 1820s when canals and barges made it possible to send the ore downriver to Baltimore and then up and down the Atlantic coast. Many of these early mines continue in operation today.

Quarries close to rivers flourished and Maryland is rich in stone suitable for building and carving. Among the most important of these stones is red Seneca sandstone. From the 1840s to 1880s brownstone buildings were popular and Maryland's sandstone floated downriver in barges to Washington, D.C. The most famous brownstone structure in the Capitol is the main building of the Smithsonian Institution - the castle. A second stone of historical significance is Cockeysville Marble. Quarried just north of Baltimore, this white, coarse-grained stone was used to build most of the Washington Monument. Maryland's Cockeysville marble formed the solid foundation, but funding dried up and construction halted for twenty-two years. Four rows of white marble from Lee, Massachusetts were laid in 1879, but the stone was too costly so the builders finished the Washington Monument with Cockeysville Marble. This later marble did not exactly match the earlier quarrying. A critical eye can distinguish the three layers of marble.

The Calvert Cliffs area is well-known to fossil collectors. Fossil deposits from the Miocene epoch lay exposed to weathering between Chesapeake Beach and Drum Point. Visitors may find fossilized shells and sharks teeth by inspecting or sifting the sand here. Sharks teeth, stingray tooth plates, coral and shells are the most common finds here, but they are not the only fossils worth hunting. Marine vertebrates such as porpoises, manatee and dolphin have been found along with bones from birds such as the booby and fulmar. On rare occasions fossils from land mammals are recovered, such as tapirs, rhinos, horses and mastodons. These creatures did not habituate this area, but perhaps rivers and floods brought their remains to the bay. This site is an exciting stop for fossil enthusiasts.

Maryland is a pleasant surprise. Her copper, gold and chromium supplied the world. White marble, and red sandstone provided the solid building blocks for much of our nation's capital. And don't overlook the barren places and cliffs where you might just find a fossilized rhino tooth or a gold nugget.

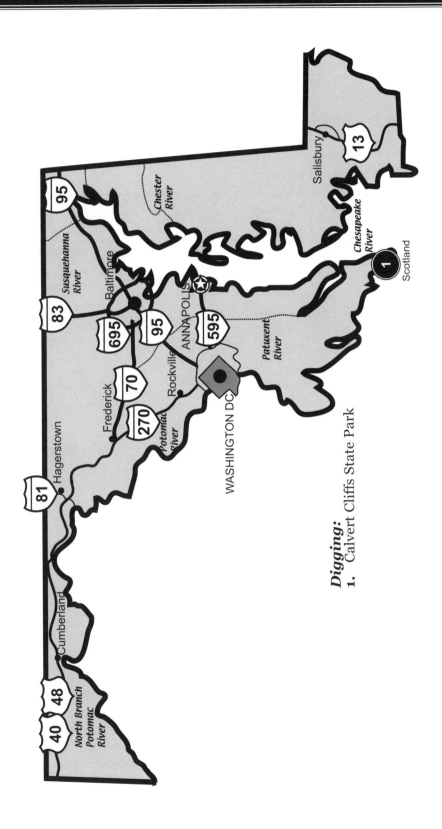

Digging:
1. Calvert Cliffs State Park

ADDRESS:
Point Lookout State Park
P.O. Box 48
Scotland, MD 20687
(301) 872-5688
http://www.dnr.state.md.us/publiclands/southern/calvertcliffs.html

DIRECTIONS:
Driving from Prince Frederick, Maryland, take State Highway 4 south approximately 14 miles. Follow the signs to Calvert Cliffs State Park. To fossil hunt, follow the signs to the park from State Highway 4, but turn left onto County Road 765, rather than into the main park entrance. Drive half a mile, just past the Chapel and turn right onto Camp Canoy Road. Continue along the paved road, bearing right when the road changes to dirt. About 0.25 of a mile along the dirt road you will come upon the entrance to Bay Breeze on the right. The combination lock code is 1224. After passing through the gate drive ahead to the parking area.

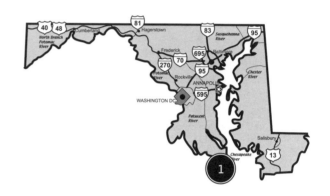

SEASON:
Open all year round

HOURS:
Sunrise to sunset – Monday through Saturday

COST:
A nominal entrance fee is charged for entrance to the park

WHAT TO BRING:
You will be hunting in the water and by the shore so wear a swimsuit or bring a change of clothes because you will get wet. Bring hand tools and a water scoop.

INFORMATION:
    Calvert Cliffs are 600 feet high and 30 miles long, rising from the western shores of the Chesapeake Bay. Generations of sea life lived and died here, sinking to the ocean floor to fossilize. These fossils are now exposed by wind and rain. Approximately six hundred species have been identified here, among them oysters, clams, scallops and sharks teeth.
    To find fossils, hike a little under 2 miles from the parking lot to the cliffs. Fossils can be found on the beach, near the shoreline and in the water. Use a water scoop to improve your chances. You may keep whatever you find. Access to the base of the cliffs is closed to hunting because of frequent landslides.

The first dinosaur footprints ever found in North America

When they first found the footprints, frozen in stone, the Moody family did not know what to make of them. The year was 1802 and dinosaurs were unknown. Although bones had been found in many places in the world. Some people even thought the giant bones belonged to dragons, which was not too far from the truth. The word dinosaur didn't exist until it was created in 1841 so it is understandable that when Pliny Moody uncovered these tracks while plowing his father's field in South Hadley, he did not know what kind of beast made the imprint. The tracks looked like those of a large bird. The family decided the prints were as old as the Great Flood, described in the Bible. They called impressions the tracks of Noah's Raven. The biblical story of Noah tells that he first sent the raven to seek land, when it did not return to the ark he sent the weaker dove. Moody used the rock of Noah's Raven's footprints as a doorstep.

In 1835 the citizens of Greenfield, Massachusetts noticed their new shale stone sidewalks bore the same strange stone turkey tracks. An expert from Amherst College was called. Professor Hitchcock collected these tracks and obtained Moody's find as well. He took them to Amherst College, where they still are today in the Pratt Museum.

Eventually these tracks became known for what they are, 225 million-year-old dinosaur tracks from the Triassic and early Jurassic periods. A wide variety of fossil prints from both biped and quadruped dinosaurs have been unearthed in this area. The latest was found in 1996 near Holyoke when a landowner was digging a hole for a fishpond. Visitors to Massachusetts who wish to see tracks where they were found or are interested in purchasing tracks can try a visit to Dino Land, a sixty-year-old local attraction in South Hadley, Massachusetts.

Ancient prints give scientists valuable clues that cannot be gleaned from fossils. They show how animals moved, how fast they moved, and where they went. Some dinosaurs were pigeon-toed, while others waddled like ducks. Prints indicate that some species moved in herds with the smaller dinosaurs toward the middle, telling scientists volumes about dinosaurs' social behavior. These impressions are truly fascinating.

Speaking of interesting rocks, Massachusetts has two very famous ones. Plymouth Rock is located just south of Boston in Plymouth, Massachusetts. This is reportedly the place where the pilgrims made their first landing in the New World. Ravaged by souvenir hunters, the rock is greatly diminished, though still holding on. This historic rock now has the date of the landing, 1620, carved upon the surface and is housed in a shelter to protect the stone from the elements and souvenir hunters carrying rock hammers.

A lesser-known treasure is the Dighton Rock, located at Dighton Rock State Park. This eleven-foot boulder once sat on the bank of the Taunton River and now resides in a museum for its protection and preservation. The stone's surface is covered with a picture language of unknown origin. The mystery of this writing keeps archeologists guessing to this day. Some believe the petroglyphs are Native American in origin. Others say they are the result of Phoenician, Portuguese or Viking explorers. In any case these markings were likely made far before Columbus ever "discovered" America and continue to fascinate all who visit this ancient boulder.

Massachusetts's final treasure has nothing to do with rocks unless you consider the ballast rock she carried. This discovery has more to do with gold doubloons and pieces of eight. The pirate ship Whydah, the famous wrecked ship of the pirate Samuel Bellamy is being salvaged in the waters off Cape Cod. There is a museum in Providence Town, which tells of the ship, pirates, and displays their ill-gotten gains.

Museums
1. Harvard Mineralogical Museum
2. Museum of Science
3. Pratt Museum of Natural History
4. The Springfield Museums at the Quadrangle
5. Whydah Museum

Points of Interest
6. Dighton Rock State Park
7. Nash Dino Land
8. Pilgrim Memorial State Park

Museums

Site 1.
Harvard Mineralogical Museum
24 Oxford Street
Cambridge, MA 02138
(617) 495-3045
http://www.peabody.harvard.edu/museum_mineral.html
Exhibits include some of the world's finest, rarest rocks, minerals, ores, meteorites, as well as rough and cut gemstones.

Site 2.
Museum of Science
1 Science Park
Boston, MA 02114
(617) 723-2500
http://www.mos.org
This museum offers many hands on exhibits that will enchant and teach kids of all ages about science. Exhibits explore sound, dinosaurs, minerals and much more. With changing exhibits and lots to do, visitors are bound to enjoy their journey of discovery.

Site 3.
Pratt Museum of Natural History
(soon to be Amherst College Geology Building and Natural History Museum)
Amherst College
Amherst, MA 01002-5000
(413) 542-2165
http://www.amherst.edu/~pratt/
This natural history museum is housed at Amherst College and includes over 10,000 specimens. This is really the second collection, as the first was all but destroyed in a fire in 1882. The Pratt Museum will be closing in 2004 while a new facility is being built. The new museum is due to open in 2006.

Site 4.
The Springfield Museums at the Quadrangle
220 State Street
Springfield, MA 01103
(413) 263-6800
http://www.quadrangle.org/
The quadrangle includes four museums including two art museums, a history museum and a science museum. The Springfield Science museum features a Dinosaur Hall with a life-size replica of Tyrannosaurus Rex. A mineral hall and hands-on exhibits investigate natural and physical science.

Site 5.
Whydah Museum
Expedition Whydah Sea-Lab & Learning Center
P.O. Box 493
16 MacMillan Wharf
Provincetown, MA 02657
(508) 487-8899
http://www.whydah.com/
The Whydah Museum is devoted to the only known pirate ship ever salvaged. Visitors learn about the recovery and view artifacts from the pirate ship captained by "Black Sam" Bellamy until it wrecked on the reef off Cape Cod in 1717.

POINTS OF INTEREST

Site 6.
Dighton Rock State Park
Bay View Avenue
Berkley, MA 01224
(508) 822-7537
http://www.state.ma.us/dem/parks/digr.htm
The petroglyphs covering an eleven-foot tall boulder draw visitors. Found on the shore of the Taunton River, the ancient carvings are now protected in a small museum which is open by appointment only. Most unusual of all is the controversy surrounding the symbols unknown origins. Theories include Portuguese explorers and Native Americans.

Site 7.
Nash Dino Land
Route 116
South Hadley, MA 01075
(413) 467-9560
http://www.roadsideamerica.com/attract/MASOUnash.html
This eclectic privately owned roadside attraction has recently re-opened and preserves many impressions made by dinosaurs who traveled the sandy ground, now stone. Some prints are available for sale.

Site 8.
Pilgrim Memorial State Park
Water Street
Plymouth, MA 02360
(508) 866-2580
http://www.state.ma.us/dem/parks/plgm.htm
This small park protects the site where the Pilgrims landed in 1620. Visitors will find Plymouth Rock, where legend has it that the Pilgrims first set foot New England. On the waterfront visitors will also find a reproduction of the Mayflower, which was the ship used to make the historic crossing.

Copper Country

Humans have mined copper in the area now known as Michigan for thousands of years. Evidence of their work remains in the open pits. Hammer stones, fire and water were used by ancient peoples to pull the rich ore from the earth. Copper artifacts traveled far and wide through trade or migration. Hammered copper knives, arrows, spearheads and axes are recovered in archeological digs. In the 1840s the United States claimed the Upper Peninsula as U.S. territory and a copper rush ensued. Miners arrived by boat as no roads reached the wild territory. In the winter, dog sleds were the transportation of choice as miners dug deeper into the earth.

Mines sprung up over the ancient open digging pits and many mining companies formed, but by the 1850s most were bankrupt after dumping thousands of dollars into unproductive shafts. The mines that survived reaped rich copper veins as this area became the world's leading copper producer. Railroads reached the Peninsula in the mid-1850s and the territory changed from wilderness to boomtowns to respectable communities. Thirteen billion pounds of copper have been extracted and 16 million ounces of silver since 1845. A huge native nugget from this area is on display in the Smithsonian Institution in Washington, D.C.

Eventually the mines became unprofitable and the companies departed leaving behind ghost towns, tailing piles, copper and mine shafts. The ore is still there in high quantity along with other precious metals.

The glaciers that shaped Michigan left gold in the gravel. These same glaciers cut into the earth exposing the existing copper veins to ancient peoples. Early geologists and surveyors discovered more than interesting landscapes. One surveyor noticed his compass deviating from true north and thus pinpointed rich iron deposits. Geologists who began the copper rush also found small quantities of gold nuggets. Unlike California, these nuggets did not lead to a mother lode farther upstream and were possibly carried by glaciers over great distances and spread far and wide. Silver and gold are here in quantities too small to make commercial mining profitable, but they supply hobbyists with a wealth of mining opportunities.

The deep mining shafts left behind rocks gleaned from deep in the earth. These tailing piles are a rockhound's dream. Tourmaline and quartz, dolomite, serpentine, jaspers and hematite are found along with copper, chrysocolla, epidote, and silver. The shores of Lake Superior yield banded agates to beachcombers. A unique stone found in Michigan is Petoskey Stone. This stone is actually fossilized coral with a unique starburst design upon the surface. Petoskey State Park is dedicated to this stone.

Those interested in the history of this area would enjoy a visit to the Upper Peninsula including Iron County Historical Museum and Coppertown USA Mining Museum. To find beach agates, try Picture Rocks National Lakeshore.

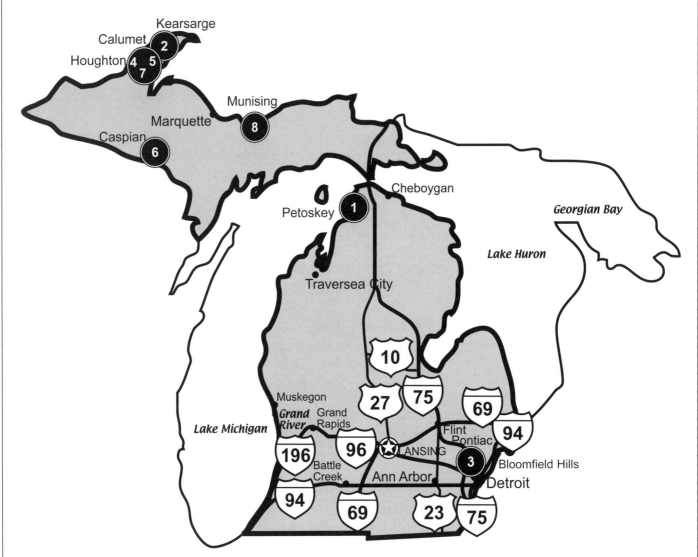

Diggings
1. Petoskey State Park–Petoskey Stone
2. Delaware Copper Mine–Copper

Museums
3. Cranbrook Institute of Science
4. A. E. Seaman Mineral Museum
5. Coppertown USA Mining Museum
6. Iron County Historical Museum

Points of Interest
7. Isle Royale National Park
8. Pictured Rocks National Lakeshore

ADDRESS:
2475 M-119
Petoskey, MI 49712
(231) 347-2311
(800) 44-PARKS for reservations
http://www.michigandnr.com/parksandtrails/ParkandTrailsInfo.asp?id=483

DIRECTIONS:
Take Interstate Highway 75 north to Indian River. Go west on State Highway 68 to Alanson. At Alanson take U.S. Highway 31 southeast to State Highway 119 north. The park is 4 miles from this junction and located on Lake Michigan.

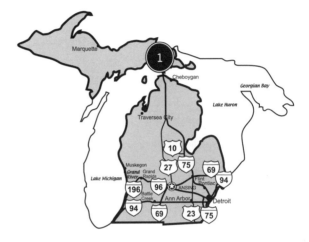

SEASON:
April 1 to October 31

HOURS:
9:00 a.m. to 5:00 p.m. (park headquarters)

COST:
Nominal admission charged by the car
$20 per night – Camping

WHAT TO BRING:
Bring a small pail to hold your stones.

INFORMATION:
Petoskey became Michigan's state stone in 1965. Petoskey stone's scientific name is Hexagonaria percarinata. This stone was once a living colony of coral on the ocean floor. It is now fossilized into a white, brown or gray stone. The striations and swirling circular pattern is what remains of the coral animal. When wet or polished, the stone shows tiny rings, resembling a honeycomb; dry it is silvery with no markings.

This fossilized coral colony weathered out of limestone and was widely distributed by glacial action. Petoskey stones are common in Michigan along the Northern and Southern Peninsulas. In the state park, they are rounded and polished by wave action and deposited on the beach.

This stone is used in jewelry, decorative items, sculpture and, of course, as paper weights. It is beautiful and relatively easy to carve. Each stone is unique.

The park is located on three hundred five acres, along Little Traverse Bay. They have a swimming area on the lakeshore. Facilities include a beach house, playground, picnic area and camping facilities. Boating and hiking are popular pastimes in this lovely park. Nearby, is the historic Gaslight Shopping District, which is worth a trip if you have the time.

ADDRESS:
Box 148
U.S. Highway 41
Kearsarge, MI 49942
(906) 289-4688
http://copperharbor.org/Business/ads/delawaremine/home.html

DIRECTIONS:
The Delaware Copper Mine is located on the Keeweenaw Peninsula and Lake Superior, south of Isle Royal National Park. From Hancock, Michigan, take U.S. Highway 41 north 43 miles to Copper Harbor. Follow the signs.

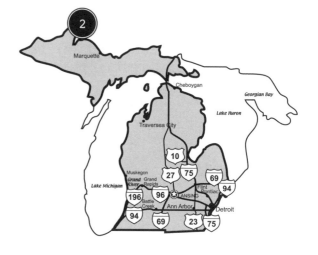

SEASON:
May to October

HOURS:
10:00 a.m. to 5:00 p.m. daily
Open until 6:00 p.m. – June, July and August

COST:
$6 – Adults
$3.50 – Children ages 6 to 12
Free – Children ages 5 and under

WHAT TO BRING:
Bring standard mining tools, your own food and plenty to drink. There is a picnic area here.

INFORMATION:
One of the oldest mines in the area, the Delaware Copper Mine began operations in 1847 and produced over 8 million pounds of copper from its five shafts. The mine closed in 1887, but is now open for tours. You will travel to the first level at a hundred ten feet below the ground and see pure veins of copper in the walls of the mine. You will learn of the area's mine history and operations.

Tours leave every twenty minutes and take forty-five minutes. The tour includes a tram ride, elevator descent and easy walking. Comfortable footwear and jackets are recommended by the mine. The temperature underground is a constant 45°F.

Also found here is a gift shop, mining museum, petting zoo and remains of mining buildings. Visitors are permitted to search for souvenirs from the mine's half-mile-long tailings pile of rock, dirt and copper. Pieces of copper range from thumbnail sized to half-a-pound. No metal detectors are permitted.

MUSEUMS

Site 3.
Cranbrook Institute of Science
39221 Woodward Avenue
Bloomfield Hills, MI 48303-0801
(877) 462.7262
http://www.cranbrook.edu/institute/
Exhibits including "Dynamic Earth" (featuring Michigan's only full size Tyrannosaurus Rex skeleton), the "Every Rock has a Story" exhibit and the "Mineral Study Gallery" will be of interest to rockhounds. Their mineral gallery exhibit includes 11,000 specimens, 300 minerals from Michigan such as gypsum and of course, copper. Precious metals and gems are displayed as well as touchable minerals.

Site 4.
A. E. Seaman Mineral Museum
Michigan Technological University
1400 Townsend Drive
Houghton, MI 49931-1295
(906) 487-2572
http://www.geo.mtu.edu/museum/
The A. E. Seaman Mineral Museum claims to have the world's finest collection of minerals from the Lake Superior Copper District with over 20,000 specimens. The displays are not limited to Michigan and include specimens from around the United States and the World.

Site 5.
Coppertown USA Mining Museum
Red Jacket Road
Calumet, MI 49913
(906) 337-4354 or (800) 338-7982
http://www.uppermichigan.com/coppertown/
At the site of the first mineral rush in the USA, the Keweenaw Peninsula, is a museum dedicated to local mining history. Beginning with Native American miners of thousands of years ago, this museum documents the history of copper mining. Exhibits are titled: Early Miners, Two-Man Drill, Foundry-Casting, The Hospital, Sheffield Pump Car and Loading Ore Cars.

Site 6.
Iron County Historical Museum
P.O. Box 272
Caspian, MI 49915
(906) 265-2617
http://www.ironcountymuseum.com/
Peer into the past at this museum, which includes many buildings from the 1890s including a homestead, lumber camp, mining area, transportation area and Victorian area. Heritage Hall contains 75 exhibits including: mining history, tool and equipment, and dioramas of mining.

POINTS OF INTEREST

Site 7.
Isle Royale National Park
800 East Lakeshore Drive
Houghton, MI 49931
(906) 482-0984
http://www.gorp.com/gorp/resource/us_national_park/mi_isle.htm
Remnants of early Native American copper mines are located here. The park is accessible by boat and sea plane only.

Site 8.
Pictured Rocks National Lakeshore
N8391 Sand Point Road
P.O. Box 40
Munising, Mi 49862.
(906) 387- 2607 or (906) 387-3700
http://www.gorp.com/gorp/resource/us_ns/mi_pictu.htm
Strata of white and yellow sandstone cliffs descend to green waters along this forty-mile picturesque shore. Many nice agates are collected here.

The iron ore turned out to be more valuable than the gold

A billion years ago the North American continent split causing iron-laden molten lava to erupt from deep within the earth. The tectonic plates pulled apart and left behind a rift called the Superior trough, now Lake Superior. This geologic event caused Minnesota to have features and resources not found in neighboring states. Water and carbon dioxide bubbled through lava and became trapped. Eventually groundwater and minerals seeped into these pockets and over time hardened into agates stronger than the lava that trapped them. Ice Age glaciers gouged into these lava beds exposing ore, agates and other riches from deep within the earth.

Native peoples found exposed copper ore. Evidence of mining includes smooth hammer stones from Lake Superior and charcoal used to heat the ore to make it more malleable. Minnesota copper was traded throughout North America. In the 1830s these ancient quarries were discovered on Isle Royal in Lake Superior. Modern miners followed the paths of the old entering burrows up to twenty feet deep. One large nugget weighed 5,720 pounds. The original weight will never be known because the copper goliath was found with charcoal surrounding a portion, and signs of hammering on the nugget. It is unknown how much ore early miners removed, but the nugget proved too large for them to remove in one piece.

Following the Civil War men discovered gold trapped in quartz and in fine grains trapped in greenstone. A rush commenced as men traveled over frozen lakes to reach Lake Vermilion in 1865. When the snow melted they found the gold scarce and difficult to recover. The rush went bust and most miners departed before the next winter snow.

The Vermilion Range had something else to offer. Some miners noticed and ignored the iron ore on their mad dash north to the gold fields. A few years later this ore became the star for Minnesota. This natural iron ore could be used without smelting. By the turn of the twentieth century, Minnesota led the nation in iron ore production, replacing Connecticut. Eventually the natural ore ran out leaving only taconite. This hard sedimentary rock must be crushed, processed and smelted to create pellets, which are shipped to steel mills. Minnesota has suffered its share of mining losses. In 1924 Minnesota lost forty-one miners at the Milford Mine when a nearby lake collapsed a tunnel wall and flooded the mine. The muddy water rose within twenty feet of ground level, trapping the miners below.

Some of the earth's riches do not require tunneling to discover them. Agates are found throughout Minnesota having weathered from the lava in which they formed. These little beauties, called Lakers by locals, have a distinct banded appearance and are commonly red, orange and yellow. Most agates are pea-sized, but some the sizes of bowling balls have been found. Agates get their color largely from iron within the quartz. Minnesota chose Lake Superior Agate as their state stone. Appropriate as this little jewel reveals the state's volcanic past and the principle mining industry – iron. Hikers and rockhounds find agates on riverbanks and lakeshores, in road cuts and streambeds. Moose Lake, Minnesota calls itself Agate Capitol of the World and annually hosts an agate festival. Activities include a rock and gem show, exhibits and an agate stampede where 150-pounds of agates and one hundred dollars in quarters in a truckload of gravel are dumped in the street for a wild scramble. Finders keepers, of course.

Minnesota has one other unique resource – Pipestone. Native Americans who used the soft material to carve pipes quarried this red stone for centuries. The mine site and the stone is and was considered sacred. The stone is believed to belong to all tribes, so no weapons are brought or used here. Pipestone National Monument preserves these quarries. Today only Native Americans can mine, stripping away the harder quartzite to reach veins of soft red stone. The rubble at the quarries reveals the labor of generations of diggers.

Minnesota's geological past is fascinating. Great forces from within the earth emerge here to give us iron, pipestone, agates suitable for jewelry, copper nuggets and gold.

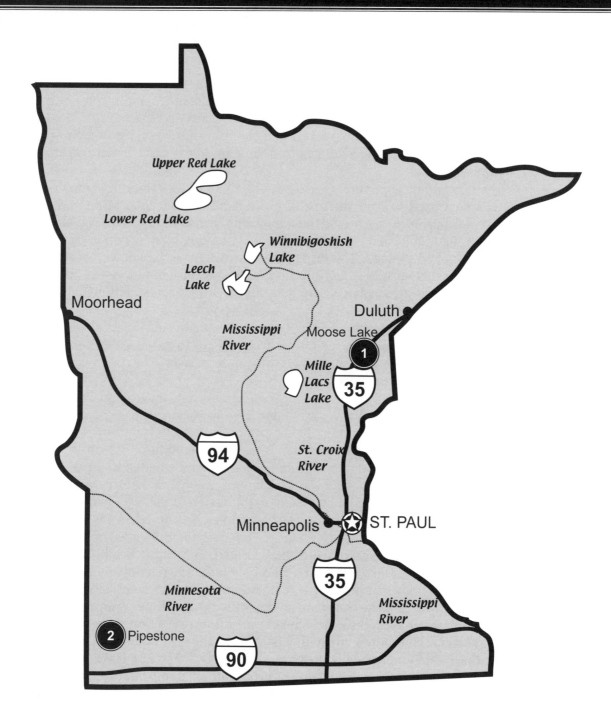

Points of Interest
1. Moose Lake
2. Pipestone National Monument

Points of Interest

Site 1.
Moose Lake
4252 County Road 137
Moose Lake, MN 55767
(218) 485-5420
http://www.mooselake-mn.com/Pages/events.html
Moose Lake, located about 100 miles north of the Twin Cities has a nice lake with swimming beach where visitors often find lovely agates. The largest agate found in the state came from Moose Lake and weighed a whopping 108 pounds! In honor of these agates, this small town hosts an annual "Moose Lake Agate Days Festival" every second full week in July.

Site 2.
Pipestone National Monument
36 Reservation Avenue
Pipestone, MN 56164-1269
(507) 825-5464
http://www.nps.gov/pipe/
Ancient peoples quarried here for catlinite, soft red stone from which they created sacred pipes.

Ziggy swam over this state 55 million years ago.

Choctaws, Chickasaws and Natchez Indians inhabited the area now called Mississippi and they were there to see Hernando De Soto, a Spanish explorer, arrive in 1540. The Choctaws allowed him and his men to winter with them before De Soto headed up the Mississippi River the following spring. In the early 1700s French settlers fought with the Natchez driving them to extinction. By the end of the century Mississippi belonged to the United States.

Mississippi is comprised of coastal areas and the Mississippi Alluvial Plain. The soil is fertile, thus Mississippi's economy is dominated by fishing and agriculture. Cotton was the prime cash crop until the boll weevil ruined this produce, now corn and soybean are also grown here. Mining natural gas, gravel, sand, cement and clay comprises only two percent of Mississippi's economy. The land is not rich in mineral wealth, but does provide gravel and sand for construction. Clay deposits are found on the shore of the Mississippi river and were deposited over 40 million years ago. The excellent plasticity makes this clay important in the ceramic industry. Clay is used primarily for floor and wall tile, next for sanitary ware, like sinks and toilets and then for pottery. One of the more bizarre uses for this clay is as an additive to animal feed to prevent the feed from caking. Mississippi has a lot of mud – they mined over one million tons of clay in the year 2000.

The Gulf Coast also contains many interesting fossils including the state fossil, petrified palmwood. Coral, sponges and mollusks from ancient beaches are often found in Mississippi. During the age of dinosaurs this state was on the bottom of the ocean. That means that land animals are seldom unearthed here, unless they traveled by water to this state. But they do have Ziggy. Ziggy is a pet name for a prehistoric carnivorous whale called Zygorhiza kochii. A skeleton of Ziggy can be seen at the Mississippi Museum of Natural History in Jackson. Ziggy wasn't too big, only about twenty feet long, which is similar to a dolphin or porpoise. Judging from the fossilized teeth he likely fed on fish and squid, though the species might have used these teeth as a strainer to gather smaller animals from the water. To know for sure you have to travel 55 million years back in time and bring a mask and fins.

MISSISSIPPI SITE MAP

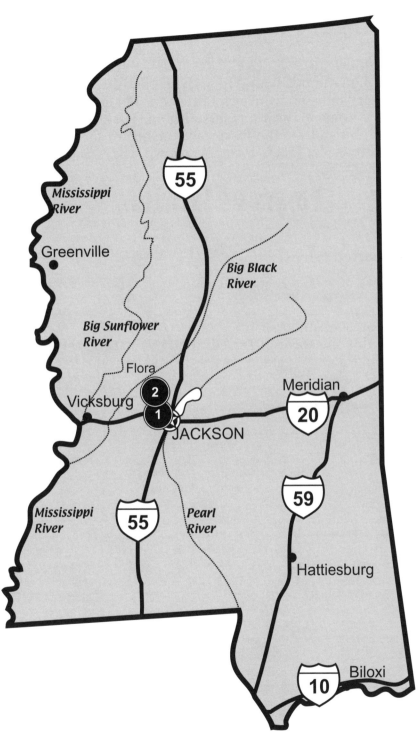

Museums
1. Mississippi Museum of Natural History
Points of Interest
2. Flora Mississippi Petrified Forest

MUSEUMS

Site 1.
Mississippi Museum of Natural History
2148 Riverside Drive
Jackson, MS 39202
(601) 354-7303
http://www.mdwfp.com/museum/
The 100,000-gallon tank aquarium is bound to grab visitors' attention. With over 200 species of fish, reptiles, amphibians and aquatic invertebrates, the display brings Mississippi's wildlife to visitors. A greenhouse is home to turtles, plants, fish and even alligators. Don't forget to visit the "Stories in Stone" exhibit and see the fossilized remains of the former inhabitants of this state including "Ziggy" the whale. Ziggy, formally known as zygorhiza, is the earliest known meat-eating species of whale.

POINTS OF INTEREST

Site 2.
Flora Mississippi Petrified Forest
P.O. Box 37
Flora, MS 39071-0037
(601) 879-8189
http://www.mspetrifiedforest.com/
Discovered in the 1800s, this site leaves petrified wood in a natural setting. Walking trails take visitors on a short, easy hike along a guided path to learn about the forces that turned wood to stone.

Colorful flint of good quality

The earliest human inhabitants of this area hunted mastodon here over 10,000 years ago as evidenced by stone spear points found in direct proximity with the giant beast's bones. Those interested prehistory would enjoy a trip to Mastodon State Historic Park in Imperial, Missouri where the Kimmswick Bone Beds have yielded fossils of extinct Ice Age creatures since the early 1800s.

These same ancient people were Missouri's first miners digging flint for arrowheads, clay for pots, pipes and bowls and iron oxide for paint.

The first Europeans to visit this area were French missionaries who came up the Mississippi and Missouri Rivers in the late 1600s. Early French explorers discovered vast deposits of lead lying everywhere on the ground in the eastern Ozarks and by 1720 the La Motte Mine was in operation.

In 1803 the United States acquired this land through the Louisiana Purchase. Thomas Jefferson commissioned Meriwether Lewis and William Clark to explore the Missouri River on their way to the Pacific. On May 14, 1804 the party left Saint Louis. Twenty-five days later they camped in a valley where the town of Rocheport now sits and discovered what the local tribes had known about for centuries – flint. Native people had long used this hard, versatile stone for the creation of tools and weapons. Lewis and Clark first saw the colorful stone on June 7, 1804. Clark commented in his journal about the red, white and blue flint of good quality and of the curious rock paintings and carvings they saw along the way. Modern visitors may see this flint and experience the beginnings of this historic trip at the Lewis & Clark State Park.

Mining for lead grew during the 1800s. The arrival of railroads greatly eased the transport difficulties experienced by many mines. In 1869 mining moved underground. This time in history is preserved at Missouri Mines State Historic Site in Park Hills. There visitors will see three galleries containing underground equipment, exhibits of lead, geology (including fluorescent minerals), and a giant air compressor used to run pneumatic drills below ground.

Missouri remains the nation's number one producer of lead. They are also first in production of refractory clay and lime, third in zinc, barite and iron, and fifth in copper and cement. These products along with crushed stone and silver are valued at over one and 500 million dollars annually.

This state has a long mining past that is still going strong. Missouri's state parks preserve areas of historic and geographic interest, and are definitely worth a look. Try Elephant Rocks State Park to see giant red granite rocks lined up like circus elephants. This same rock had been mined commercially since the 1800s and used for building and polished for monuments.

Missouri has an extensive collection of show caves, some with very interesting histories, and of course there is that very beautiful flint.

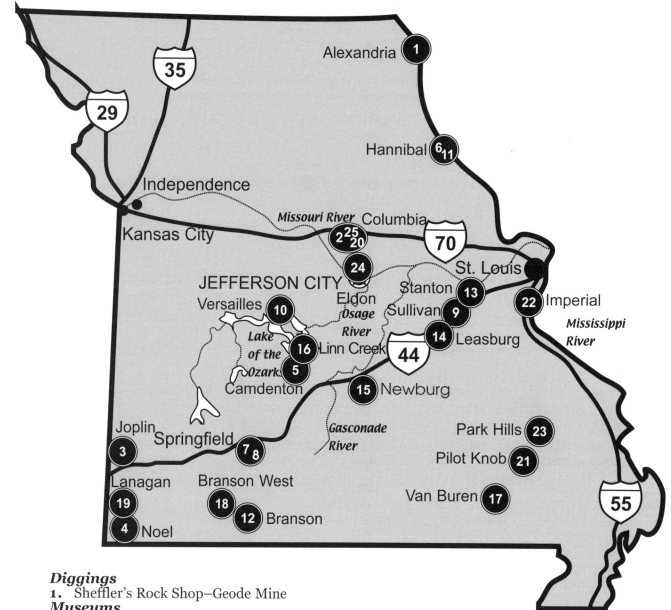

Diggings
1. Sheffler's Rock Shop–Geode Mine

Museums
2. Enns Entomology Museum
3. Tri-State Mineral Museum

Caves
4. Bluff Dweller's Cave
5. Bridal Cave
6. Cameron Cave
7. Crystal Cave
8. Fantastic Caverns
9. Fisher's Cave Meramec State Park
10. Jacob's Cave
11. Mark Twain Cave
12. Marvel Cave
13. Meramec Caverns
14. Onondaga Cave and Cathedral Cave

15. Onyx Mountain Caverns
16. Ozark Caverns
17. Round Spring Caverns
18. Talking Rock Cavern
19. Truitt's Cave
20. Wild Cave Tour Reservation

Points of Interest
21. Elephant Rocks State Park
22. Mastodon State Park
23. Missouri Mines State Historic Site
24. Rocheport
25. Rock Bridge Memorial State Park

ADDRESS:
Rural Route 1, Box 172
Alexandria, MO 63430
(660) 754-6443

DIRECTIONS:
The Sheffler's Rock Shop and Geode Mine is
located on U.S. Highway 61 between Canton,
Missouri and Keokuk, Iowa.

SEASON:
April 1 to December 1 – Weather permitting

HOURS:
9:00 a.m. to 5:00 p.m. – Open 7 days a week

COST:
$10 per person per day – Fee

WHAT TO BRING:
Bring standard mining tools, your own food
and plenty to drink.

INFORMATION:

Sheffler's Rock Shop and Geode Mine is located in northeast Missouri, close to the Illinois
and Iowa borders. The geode mine is a safe and easy way for the whole family to have a
rockhounding adventure. Stop at the rock shop first to arrange to dig and to see the display of
opened geodes from the mine. The shop has minerals for sale including rough cut rock, agates,
slabs, supplies and jewelry mountings.

Geodes are bubbles of rocks that resemble stone snowballs. They are formed by volcanic
action. Although the exterior is quite plain, the interior can be spectacular. They are
occasionally hollow and filled with crystals. The weight of the geode may indicate to you if it is
hollow. The trick, of course, is getting into them. You will need a rock saw or access to one to
see if you have a unique specimen.

The mine allows you fifty pounds of material for each admission ticket. If you collect over
fifty pounds of geodes, you pay $.50 per pound. Bring your own lunch and have a great day
looking for geodes.

Museums

Site 2.
Enns Entomology Museum
University of Missouri
Room 3-37, Agriculture Building
Columbia, MO 65211
(573) 882-2410
http://www.museum.insecta.missouri.edu/
If you or your children love bugs this is the place to be. Founded in 1874, this collection holds millions of specimens including over 600 fossil samples.

Site 3.
Tri-State Mineral Museum
7th Street and Schifferdecker Avenue
Joplin, MO 64801
(417) 623-2341
Local minerals, a scale model of a lead mine, and antique mining equipment illustrate the workings of the state's mining history.

Caves

Site 4.
Bluff Dweller's Cave
Route 2, Box 230
Noel, MO 64854
(417) 475-3666
http://www.4noel.com/bluffd
Tours take about forty-five minutes as visitors travel past various types of cave formations. The museum gift shop contains rocks, minerals, arrowheads and fossils.

Site 5.
Bridal Cave
526 Bridal Cave Road
Route 2, Box 255
Camdenton, MO 65020
(573) 346-2676
http://www.bridalcave.com
Bridal Cave gains its name from the legend of an Osage Indian wedding that may have been performed here. Brides are still welcome and over 1800 couples have wed here in a stalactite adorned chamber. Recent exploration has discovered new chambers and a subterranean lake.

Site 6.
Cameron Cave
Highway 79 South
P.O. Box 913
Hannibal, MO 63401
(573) 221-1656
http://www.marktwaincave.com/CaveHistory.html
This newest show cave in the state was discovered when steam was spotted coming from the ground one winter in 1925. Tours are conducted by lantern.

Site 7.
Crystal Cave
7225 N. Crystal Cave Lane
Springfield, MO 65803
(417) 833-9599
http://wcnet.net/adc/Cave.htm
Owners have made every effort to keep this cave, opened in 1893, in its natural state. Enjoy the underground waterfall and unique formations.

Site 8.
Fantastic Caverns
4872 N. Farm Road 125
Springfield, MO 65803
(417) 833-2010
http://www.fantasticcaverns.com
Tour this cave in a jeep-drawn tram through caverns and past unique formations.

Site 9.
Fisher Cave
Meramac State Park
2800 S. Highway 185
Sullivan, MO 63080
(573) 468-6072
http://www.umsl.edu/%7Ejoellaws/ozark_caving/comcaves/fisher.htm
Open April through October, Fisher Cave provides naturalist guides who tour visitors by lantern light through spacious chambers.

Site 10.
Jacob's Cave
Route 2 Box 129
Versailles, MO 65084
(573) 378-4374
http://www.jacobscave.com
When the cave opened 1932, visitors toured along wooden planks by lantern light. Prehistoric bones of mastodon, bear and peccary have all been found in this cave along with the world's largest known geode. The mile-long tour takes visitors past soda-straw formations and stalactite columns. See evidence of six Ice Ages and three earthquakes recorded in the living history of this cave.

Site 11.
Mark Twain Cave
Highway 79 South
P.O. Box 913
Hannibal, MO 63401
(573) 221-1656
http://www.marktwaincave.com
Samuel Clemens (Mark Twain) did not discover this cave, but he did explore it as a boy and no doubt used his childhood experiences in writing Tom Sawyer. This is the oldest show cave in Missouri and once was used by Indians, trappers, slaves traveling the Underground Railroad and the infamous Jesse James.

Site 12.
Marvel Cave
H.C.R. 1 Box 791
Silver Dollar City
Branson, MO 65616
(417) 338-8220 or (800)-952-6626
http://www.umsl.edu/~joellaws/ozark_caving/comcaves/marvel.htm
The 1880s reproduction mining town called, Silver Dollar City, has a show cave known by the Osage Indians since the 1550s. Once the Ozark Jubilee music show was held in the cave's giant chambers. Other strange happenings include a hot air balloon, which lifted off in the twenty-story main chamber to celebrate the cave's 100th anniversary of tours.

Site 13.
Meramec Caverns
Interstate Highway 44 Exit 230
Stanton, MO 63079
(573) 468-3166/(800) 334-6946
http://www.americascave.com
Guided tours along well-lit passages take visitors past limestone formations, which took thousands of years to grow. This cave boasts some of the rarest and largest cave formations in the world. Visitors throughout the cave's long history include Indians, European miners, explorers, Civil War soldiers, escaped slaves and train robbers. Open all year round.

Site 14.
Onondaga Cave and Cathedral Cave
7556 Highway H
Leasburg, MO 65535
(573) 245-6600
http://www.mostateparks.com/onondaga.htm
This state park contains Onondaga and Cathedral Caves. Listed as a National Natural Landmark, this site contains very beautiful formations and colorful deposits.

Site 15.
Onyx Mountain Caverns
14705 - P.D. 8541
Newburg, MO 65550
(573) 762-3341/(573) 762-2449
http://www.pulaskicountyweb.com/onyxcave/
Beautiful formations and natural scenic wonders combine with local history in this unique tour. Once used by Indians for shelter, flint artifacts indicate the cavern may also have been used for ceremonies. The huge natural entrance rises forty feet high and one hundred feet long. In the 1800s, cartloads of onyx were removed from the cave for use in Victorian homes as decorative stone. Beside the underground river is evidence of a "bear bed" where bears once hibernated.

Site 16.
Ozark Caverns
Lake Ozark State Park
State Road "A"
Linn Creek, MO 65052
(573) 346-2500
http://www.umsl.edu/~joellaws/ozark_caving/comcaves/ozcavrn.htm
This state park contains a large, shallow cave shown since 1952. Park staff offers tours from March through October.

Site 17.
Round Spring Caverns
Ozark National Scenic Riverways
National Park Service
P.O. Box 490
Van Buren, MO 63965
(573) 323-4236 or (573) 858-3297 ext. 23
http://www.nps.gov/ozar/caves.html
Formations of helictites, spathites, rimstone dams, stalactites and stalagmites cover this cave in formations of varying size and color.

Site 18.
Talking Rock Cavern
423 Fairy Cave Lane
Branson West, MO 65737
(417) 272-3366/(800) 600-CAVE
http://www.talkingrockscavern.com/
This cavern has many lovely crystals that are still growing. These living formations grow in different colors and textures, sometime right beside each other. Tours take about one hour.

Site 19.
Truitt's Cave
P. O. Box 190
Lanagan, MO 64847
(417) 436-2299
http://www.umsl.edu/~joellaws/ozark_caving/comcaves/truitts.htm
This cave contains rare formations and endangered salamanders. After finishing the tour, including the underground river, visitors can enjoy the activities aboveground. Visit the rock shop, walk beside waterfalls, or shop at the frontier flea market. Bring a lunch to enjoy the picnic area.

Site 20.
Wild Cave Tour Reservation
Rock Bridge Memorial State Park
5901 South Highway 163
Columbia, MO 65203
(573) 449-7402
http://www.mostateparks.com/rockbridge/cave.htm
Devil's Icebox Cave can be explored on park-lead excursions by advanced booking only. This tour takes strength and stamina. A forty-five minute orientation prepares group members who will each carry a sixty-five pound canoe one-quarter mile to the cave entrance before paddling into the cave and then scrambling through passages and up banks. This trip takes up to twelve hours.

Points of Interest

Site 21.
Elephant Rocks State Park
c/o Fort Davidson State Historic Site
P.O. Box 509
Pilot Knob, MO 63663
(573) 546-3454/(800) 334-6946
http://www.mostateparks.com/elephantrock/geninfo.htm
This park features giant boulders of granite over 1.5 billion years old standing in a row like elephants in a circus. The rocks exist as a reminder that all but the strongest stone eventually erodes away. Elephants are still appearing from the ground. Dumbo is one of the most popular boulders in the park and stands twenty-seven feet tall. Just beyond the park is the state's oldest commercial granite quarry opened in 1869.

Site 22.
Mastodon State Park
The Kimmswick Bone Bed
1050 Museum Drive
Imperial, MO 63052
(800) 334-6946
http://www.mostateparks.com/mastodon/geninfo.htm
This state historic site preserves an important archaeological find – the Kimmswick Bone Bed, in which was found evidence that humans and mastodons coexisted. During the Ice Age giant ground sloths, mastodons and other odd creatures lived the Midwest. This area may have been a swamp that trapped animals and preserved their skeletons. Visitors can hike a trail to where the bones were found, then visit the museum to learn more.

Site 23.
Missouri Mines State Historic Site
Highway 32
P.O. Box 492
Park Hills, MO 63601
(573) 431-6226/(800) 334-6946
http://www.mostateparks.com/momines.htm
This area is called the Old Lead Belt, because it once was the foremost lead mining region in the world. The St. Joseph Lead Company donated twenty-five buildings to the state for preservation in 1975 and the land is now a historic site. The old powerhouse contains a museum providing exhibits and interpretive programs to visitors. Mining equipment and related materials are on display. The museum shop sells gifts, jewelry and books related to minerals and mining.

Site 24.
Rocheport
P.O. Box 176
100 Jefferson Street
Lohman Building, Suite 200
Jefferson City, MO 65102
(573) 522-9019
http://www.lewisandclark.state.mo.us/Rocheport.asp?region=7
Captain Clark liked the colorful flint found by a nearby stream so much he noted it in his journal on his journey west. Today the town of Rocheport is proud of its historic past and provides visitors with a combination of history, geology and small town charm. Nearby attractions include the Katy Trail, Lewis and Clark Trail and the Lewis and Clark Cave. Small Native American petroglyphs are visible from Katy Trail.

Site 25.
Rock Bridge Memorial State Park
5901 South Highway 163
Columbia, MO 65203
(573) 449-7402/(800) 334-6946
http://www.mostateparks.com/rockbridge/geninfo.htm
Visitors can travel the half-mile trail across a rock bridge to the Devil's Icebox to see the underground stream and sinkhole. Above ground there are fifteen miles of trails open to the public for hiking, horseback riding and biking. Connor's cave may be explored on your own.

Books of mica smuggled out under cover of darkness

In Colonial times glass windows were imported at colossal expense. Many homes used greased paper over window openings instead. The law required that no products be manufactured in the Colonies and that all goods be imported. This is why the discovery of large transparent books of mica, also called isinglass, was so important. The mica crystals were split and used as window coverings. Mica's heat resistance made it ideal for lantern panes and windows in stoves. In 1803 a farmer named Sam Ruggles from Grafton, New Hampshire discovered a large deposit of mica and feldspar. He wisely kept this discovery secret and the mica was smuggled to Boston with his produce. Once there, Ruggles had relatives in the shipping business that sold the mica for him. Ruggles did not want American buyers to discover his secret supply of mica. Soon demand for mica was more than he could fill. He hired workers, swearing them to secrecy. The mica was transported to Boston under the cover of darkness by horse and wagon.

Word of the mica discovery did eventually leak out. The Ruggles mine is now open to the public. Visitors can find white and black mica, feldspar and various other rocks and gems associated with feldspar including: amethyst and beryl.

Mica is still mined and used in various products. This stone gives some lipstick and eye shadow its glimmer and adds viscosity and toughness to exterior paint. It is used in highway construction, caulk, lubricants, grease, and in dry-powder fire extinguishers.

The feldspar mining began here in 1825. In 1886 J.T. Robertson Soap Company purchased large quantities of quartz from the Ruggles mine in order to make a scouring soap from ground quartz and tallow. The quartz came with feldspar stuck all over it and this had to be broken away with hammers. The annoying substance was discarded until someone decided to use it as a non-abrasive cleaner. The feldspar was ground into fine powder mixed with liquid soap and dried into cakes. Bon Ami was sold for ten cents a cake. The packaging included a fluffy yellow chick and the slogan "hasn't scratched yet." The meaning of this was apparent to farmers who used the product. They all knew that newly hatched chicks live several days off the yoke from the egg, and do not scratch at the dirt. This slogan is lost on modern users, but Bon Ami survives and is still a useful cleaning product.

Associated with the feldspar were crystals of green and black. These were discovered to be beryl including: aquamarine, emerald, heliodor and morganite. These gems were mined and sold for jewelry and were New England's best known early source of gems. During WWII these gems found other uses in electrical devices used in aviation and missilery.

Not all the quartz found here was ground into abrasive cleaners. Rockhounds can collect specimens including clear, smoky and amethyst. One more gem is found in New Hampshire, a lovely clear crystal called topaz.

Mica, feldspar, quartz and an abundance of gemstones, that is more than enough for New Hampshire to earn a place in the heart of rockhounds everywhere. But New Hampshire is also famous for the lovely white granite quarried near Rattlesnake Hill. It is this stone that gives New Hampshire its nickname "the granite state." The stone quarried here was used to build the Library of Congress in Washington, D.C. among others. The quarries are still active, producing stone. If you look closely at buildings, monuments and cemeteries throughout New England, you will see this state's famous granite hard at work.

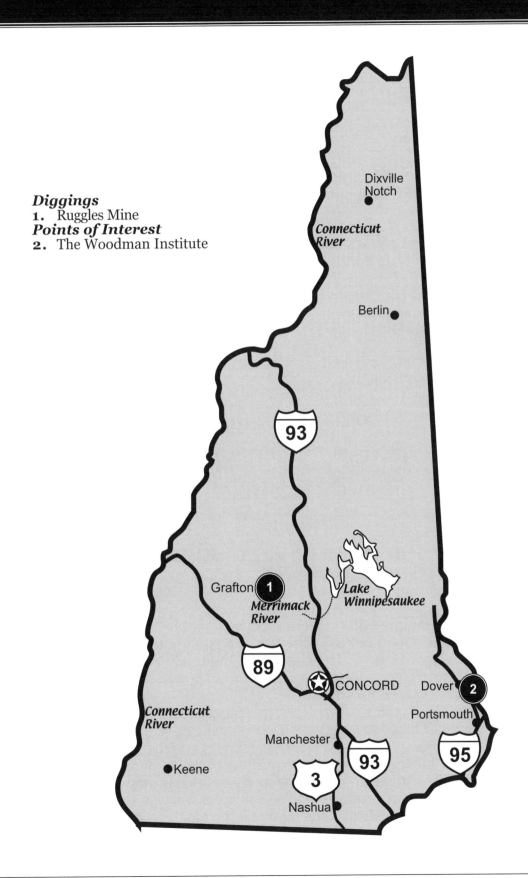

Diggings
1. Ruggles Mine
Points of Interest
2. The Woodman Institute

Dixville
Notch

*Connecticut
River*

Berlin

93

Grafton **1**
*Merrimack
River*

Lake
Winnipesaukee

89

CONCORD   Dover   **2**

Portsmouth

*Connecticut
River*

Manchester

93   95

Keene

3

Nashua

ADDRESS:
Route 4 at the Village Green
Grafton, NH 03240
(603) 523-4275
http://www.rugglesmine.com

DIRECTIONS:
Ruggles Mine is north and west of Grafton, New Hampshire. Take U.S. Highway 4 west approximately 9 miles, following the signs. Turn onto the mine access road at the sign across from the white church. The access road is unpaved.

SEASON:
Open weekends – mid-May to mid-June
Open weekdays – mid-June to mid-October

HOURS:
9:00 a.m. to 5:00 p.m. – May and June
9:00 a.m. to 6:00 p.m. – July and August
The last ticket sold one hour before closing

COST:
$15 – Adults (this price buys all the rocks you can haul to your car in two trips)
$7 – Children ages 4 to 11

WHAT TO BRING:
No tools are necessary to find specimens. You may wish to have a rock hammer and bucket. Some supplies can be rented at the mine. If you want to explore the shallow caves, bring a flashlight. A snack bar and picnic tables are available.

INFORMATION:
This mine boasts that one hundred fifty minerals can be found here. The most common are mica, feldspar and beryl. This mine is actually the top of a mountain. They suggest you leave trailers at the bottom of the gravel road. The drive up is definitely worth the trip. There are spectacular views of the surrounding mountain. The entrance cuts through the mountain, leading you into the quarry. The entire top of the mountain has been mined away and the interior is open to the sky. High pillars of feldspar rise on each side of the entrance like the columns of a Roman temple. There are some small caves that can be explored from the main mining area.

Mica was coveted for its transparence and fire-resistant qualities, and was commonly used for lanterns and stove windows. It is still used today in the cosmetics industry to make lipstick glitter. Feldspar is also used in many products, as you will see if you visit the mine. Visitors can take a walking tour of the mine's history, starting at the outside the mine's entrance in the museum.

POINTS OF INTEREST

Site 2.
The Woodman Institute
182 Central Avenue
Po Box 146
Dover, NH 03820
(603) 742-1038
http://www.seacoastnh.com/woodman/
This museum contains rocks, minerals, Native American artifacts and other natural history exhibits.

Messages from the past, trapped in liquid stone.

Millions of years ago, much of New Jersey lay submerged beneath the ocean. As waters receded, plants and land animals pervaded. Rotting organic became a peaty black material and was pressed by time and pressure into coal. Some of this black organic matter became carbonized into lignite and was trapped in clay. Much of the modern landscape was shaped during the Ice Age by advancing and retreating glaciers. Lakes were gouged and bedrock scraped clean in places, exposing outcroppings of iron ore.

Modern man discovered the outcroppings of iron ore and began mining. As early as 1731 furnaces belched black smoke into the sky. In the 1760s a village called Batsto rose on a river southeast of Philadelphia. The waterways gave easy access to markets, and the surrounding riverbanks and swamps were rich with bog iron created from the oxidation of vegetative matter. The Batsto Iron Works supplied Washington's Continental Army with cannons, cannonballs, wagon parts, ship parts and kettles. When richer deposits of iron were found in Westbrook Valley, mining operations were forced to change. Batsto quickly switched from smelting iron to the production of glass. Great quantities of fine sand and smelting furnaces provided all they needed. The state of New Jersey has preserved Batsto Historic Village within the Wharton State Forest. For information about New Jersey's glass making history visit the Heritage Glass Museum in a town aptly named Glassboro.

While coast mining floundered, inland mines flourished. Shafts were dug deep into rich veins of ore exposed by glaciers. Zinc was mined in the Franklin-Sterling Hill area. These mines operated for two centuries. The last closed in the 1980s. New Jersey has preserved her mining past in museums. Fantastic displays of fluorescent minerals, exhibits highlighting local mining history, and examples of the 300 plus minerals discovered here are found at the Franklin and Sterling Hill Museums.

Amber found in central New Jersey is over 92 million years old. Some of the most scientifically significant amber in the world was found here, including the earliest flowers found: oak tree flowers. Scientists studying prehistoric plants usually glean information from fossil impressions, but amber traps the actual plant, preserving it intact. Another unique species found in amber includes a moth whose feeding parts indicate the species was in the process of transforming from a biting to a nectar-feeding insect. Mosquito, black fly and other biting insects may contain DNA from the dinosaurs they bit.

Significant fossils have also been found in New Jersey, and it is where the infamous Bone Wars between paleontologists Edward Drinker Cope and Othniel Marsh erupted. Most fossils recovered here are the remains of marine animals, since the area was below sea level during the age of dinosaurs. Construction sites are likely places to find fossilized clams, oysters, shark teeth and coral. On rare occasions a crab or lobster claw is unearthed. Even rarer are turtle shells, crocodile skeletons or the remains of bony fish. A truly unique discovery was made in 1838 when workers digging a mixture of clay and lime for fertilizer uncovered some large bones. Twenty years later scientist William Foulke heard of the bones discovered in the marl pits and resumed digging. What he found was the first complete dinosaur skeleton ever found in the world. The skeleton's reconstruction changed thinking that all dinosaurs were quadrupeds. The huge duck-billed dinosaur, called Hardrasaurus foulkii, was displayed in 1868 at the Philadelphia Academy of Science where the fossil is still on display today. Philadelphia became a scientific Mecca for paleontologists and New Jersey's marl pits supplied them all. Enter Edward Cope, a prominent scientist who paid marl workers to supply him with any recovered bones. He shared this information with a colleague named Othniel Marsh. The relationship turned ugly when Cope discovered that Marsh paid these same marl workers more to send him the bones. These events set up an intense rivalry that moved west with these men as they raced to discover and name new species, and is known as the Bone Wars.

The discovery of new fossils in New Jersey declined as the marl pits closed. Greater discoveries were made in Wyoming and the search moved west. But for those who know where to look, fossils, amber and fabulous florescent minerals are still here where it all began – in New Jersey.

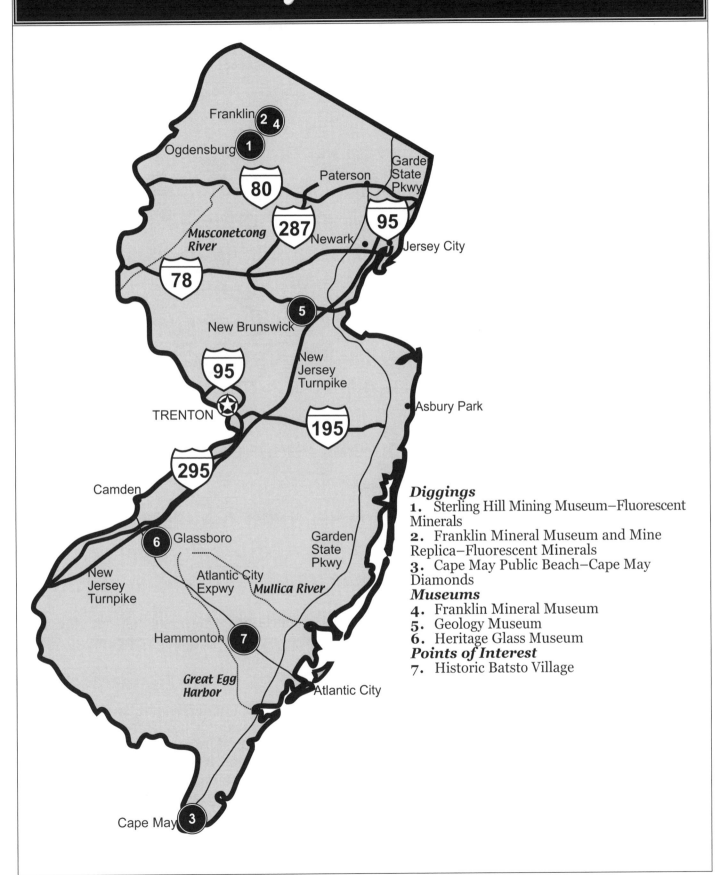

Diggings
1. Sterling Hill Mining Museum–Fluorescent Minerals
2. Franklin Mineral Museum and Mine Replica–Fluorescent Minerals
3. Cape May Public Beach–Cape May Diamonds

Museums
4. Franklin Mineral Museum
5. Geology Museum
6. Heritage Glass Museum

Points of Interest
7. Historic Batsto Village

ADDRESS:
30 Plant Street
Ogdensburg, NJ 07439
(973) 209-7212
http://www.sterlinghill.org/

DIRECTIONS:
The Sterling Hill Mine is located a short distance from either Sparta or Franklin, New Jersey off of County Road 517. Take County Road 517 to Ogdensburg and turn north onto Passaic Avenue, driving 0.7 mile to the entrance to the mine.

SEASON:
Collecting is allowed the last Sunday of each month during the operating season
March 1 to November 30 daily (tours)

HOURS:
10:00 a.m. to 5:00 p.m.
Tours leave at 1:00 p.m. and 3:00 p.m.

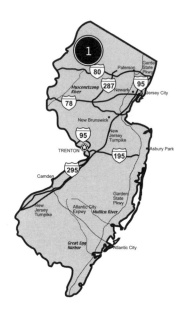

COST:
Mineral Collecting
$10 per person - must be 14 years of age or older

Mine Tour
$9.50 – Adults
$8.50 – Seniors
$7.00 – Children under 12 years of age

WHAT TO BRING:
No special equipment is needed for the tour. For Sunday collecting, bring small buckets, gloves, safety goggles, rugged footwear and a rock hammer. An ultraviolet light can be helpful to see which rocks fluoresce.

INFORMATION:
Sterling Hill is the last operating zinc mine in New Jersey with underground, ancient mine pits and antique mining equipment. The mine museum has an exhibit hall featuring mining artifacts and rare minerals. Walking tours cover 0.25 of a mile of underground tunnels and last between half an hour to two hours. This method of metal mining is explained as you view tunnels, tools and equipment. Bring a sweater or jacket, as the tunnels are a cool 52°F year round. The highlight of the tour is seeing the spectacular red and green minerals fluoresce within the mining tunnels.

Please call and check on the time and price of collecting. Children must be 13 years of age or older to collect at the mine.

ADDRESS:
32 Evans Street
Franklin, NJ 07416
(973) 827-3481
http://www.franklinmineralmuseum.com

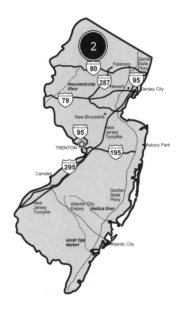

DIRECTIONS:
The Franklin Mineral Museum is located off State Highway 23. From State Highway 23 take Route 631 and follow the signs to the museum.

Seasons:
April 1 to December 1 – closed Easter and Thanksgiving
March (weekends only) – Weather permitting

HOURS:
10:00 a.m. to 4:00 p.m. – Monday through Saturday
11:00 to 4:30 p.m. – Sunday

COST:
$6 – Adult tour fee
$3 – Child tour fee
$4 - Senior tour fee
$11 – Adult tour and collecting fee
$6 – Child tour and collecting fee
$7 - Senior tour and collecting fee

WHAT TO BRING:
Bring standard mining tools, your own food and plenty to drink. An ultraviolet light might be helpful, although one is available at the museum.

INFORMATION:
The mineral museum is outstanding with over 4,200 specimens on display. The collection of fluorescent minerals is very impressive. Take note of how the minerals look in regular light so you can try to locate them later in the tailings pile. There is also a collection of fossils, petrified wood, shells and Native American artifacts for your enjoyment.

Franklin is the site of the former New Jersey Zinc Company. The replica mine displays the equipment and methods used to extract zinc. Collecting is done at the Buckwheat Tailings next to the museum. Reservations must be made in advance for group tours.

ADDRESS:
County Route 606
Cape May, NJ 08204
(609) 884-9562 (Cape May Welcome Center)

DIRECTIONS:
Take the Garden State Parkway south to its end. Follow the signs to Cape May on County Road 606.

SEASON:
Open all year round

HOURS:
Daylight hours

COST:
Free

WHAT TO BRING:
Some beach toys, a closed container and a sharp eye are all that you will need.

INFORMATION:
Cape May Diamonds are small quartz pebbles that have washed up on the beach. They can be found on several beaches in the area including Lighthouse State Park, Cape May Point and Sunset Beach. Look for transparent stones at these sites. For the best results, tumble the stones for three weeks. They can be cut by a lapidary to look just like diamonds. Visit local gift shops to see what can be done with these lovely stones.

Cape May is a pretty Northeast summer getaway. The town is just south of Atlantic City and Wildwood, New Jersey. This village has a different look than most resort towns—no high-rises. Guesthouses with gingerbread wood trim prevail. The shopping area is touristy, but nice. From Sunset Beach you will see what looks like the remains of a cement building in the ocean. It is actually an experimental cement ship that was constructed during World War II. The ship sailed, but ran aground in a storm and remains where it struck the bottom.

If you tire of the peace and quiet of Cape May, Wildwood is a hop, skip and a jump away. This summer mecca has a large boardwalk, which has carnival-style rides, food, shops and fun. Lots of hustle, bustle, and action to be found here, along with a very nice beach. Bring your metal detector!

MUSEUMS

Site 4.
Franklin Mineral Museum
32 Evans Street
P.O. Box 54
Franklin, NJ 07416
(973) 827 3481
http://www.franklinmineralmuseum.com/index.htm
This museum displays 300 varieties of minerals from the Franklin-Ogdensburg area, including polished gemstones, colorful crystals, and a large collection of fluorescent minerals. Also on display are collections of fossils, Native American relics, and a life-size replica of a zinc mine.

Site 5.
Geology Museum
Rutgers University
Geology Hall/Old Queens Section
College Avenue Campus
New Brunswick, NJ 08903
(732) 932-7243
http://urrutgers.edu/geology/
This university museum contains displays of geology and anthropology that emphasize the natural history of New Jersey. The largest exhibits include a dinosaur trackway and a mastodon, from New Jersey.

Site 6.
Heritage Glass Museum
High and Center Streets
Glassboro, NJ 08028
(856) 881-7468
http://www.fieldtrip.com/nj/98817468.htm
Glassboro gains its name from the town's glass-making history. At the Heritage Glass Museum the past is preserved. Beginning in 1779, glass was made and fashioned into bottles, pane glass and other items, which are now displayed in exhibits here.

POINTS OF INTEREST

Site 7.
Historic Batsto Village
Wharton State Park
4110 Nesco Road
Hammonton, NJ 08037
(609) 561-0024
http://www.batstovillage.org/
This village is part of Wharton State Forest, but in the 1700s this was a bustling, prosperous town. Residents made iron and glass using local resources. The iron was important to America's Revolutionary War efforts. This village offers yearly special events and allows visitors to tour buildings and see a slice of early American life. Tours include the grist mill, water-powered wood mill, general store, post office and the mansion. There is a museum shop on site.

Double terminated clear crystals, unique in all the world.

A glacier vastly changed New York's topography and flattened some of the oldest mountains in the country, the Adirondacks. These mountains made mineralogical history in the late 1800's when garnets were discovered on Gore Mountain in North Creek by Henry Hudson Barton, Sr., a jeweler's assistant. He founded a mine in 1878, and the business has been going ever since, making sandpaper (garnet paper) and other abrasives. Garnets became the official state mineral.

Garnets are found in other places in the world, but one of New York's treasures is unique in all the world. Herkimer Diamonds are found only in the area in or near the town of Herkimer, New York. These quartz crystals have the distinction of forming in perfect, clear, double-terminated shapes with eighteen facets each. These crystals grew over five million years ago. Iroquois and Mohawk Indians first found them and used the quartz to fashion tools and weapons. Local folklore claims the quartz crystals were later discovered by two soldiers during the American Revolutionary War, who brought them to the attention of General Herkimer. According to the tale, the General believed them to be real diamonds and had high hopes of financing the war with the gems, but reports from Europe identified them as merely quartz, if an extremely rare formation.

Both garnets and Herkimer diamonds are still abundant for those who know where and how to look. Barton Mine gives tours and allows visitors time to collect samples. The town of Herkimer has two mines, the Herkimer Diamond Mine and the Ace of Diamonds. Outside Herkimer are two more mines: Treasure Mountain in Little Falls, and Crystal Grove in Saint Johnsville. The Little Falls diamonds have the distinct feature of forming in scepters, which is one larger clear stone atop one or more longer darker stones. Such formations are rare and highly prized. Visitors to the mines use hand tools to break rocks searching for hollow pockets containing diamonds. More ambitious miners use sledgehammers and metal wedges to take down the stone wall in search of pockets in the stone. Such pockets, called vugs, are known to hide the biggest diamonds and clusters of crystals. Those not wishing to break boulders can search the rock piles for diamonds or keep an eye to the ground for glittering quartz.

The state ranks 15th in the nation in mineral production with over $1.5 billion in mined products recovered per year. Some of the more utilitarian products mined here include limestone for cement, clay for brick and sand, gravel and crushed stone for building. Lead and silver are mined in association with zinc. Granite, slate and sandstone are also quarried in New York. New York is the only place in the U.S. that mines wollastonite. This mineral has a needle-like structure and is used for cleaning things such as car bumpers, vinyl floors and teeth! New York also commercially mines salt, zinc and talc for such products as sandpaper, wallboard, brick, glass, paint pigments and abrasives.

Fossils ranging in size from tiny seashells to mastodon bones are prevalent in New York. The Buffalo Museum of Science sponsors yearly expeditions to unearth new finds. For the general public, Treasure Mountain in Little Falls has an excellent shale collecting site where visitors can split shale in search of fossil impressions of trilobites, snails, worms and plants.

New York has some of the best museums in the world. Rockhounds will be impressed with Buffalo's Museum of Science. Those interested in rocks and minerals indigenous to New York State should visit the New York State Museum in Albany. The crown jewel of museums has to be New York City's American Museum of Natural History, which houses vast permanent exhibitions in the Hall of Gems and Minerals. Gem enthusiasts will see one of the world's most famous sapphires, the Star of India, along with a huge collection of other precious gems. Gold nuggets and topaz weighed in pounds are impressive. The museum now has a new wing dedicated to space, which houses meteors of colossal mass. One can easily spend an entire day in the mineral hall. Don't forget to visit the dinosaurs. The American Museum of Natural History has some of the most rare and complete fossils ever recovered.

So whether you wish to dig your own or see what others have recovered, New York has what you're looking for.

NEW YORK SITE MAP

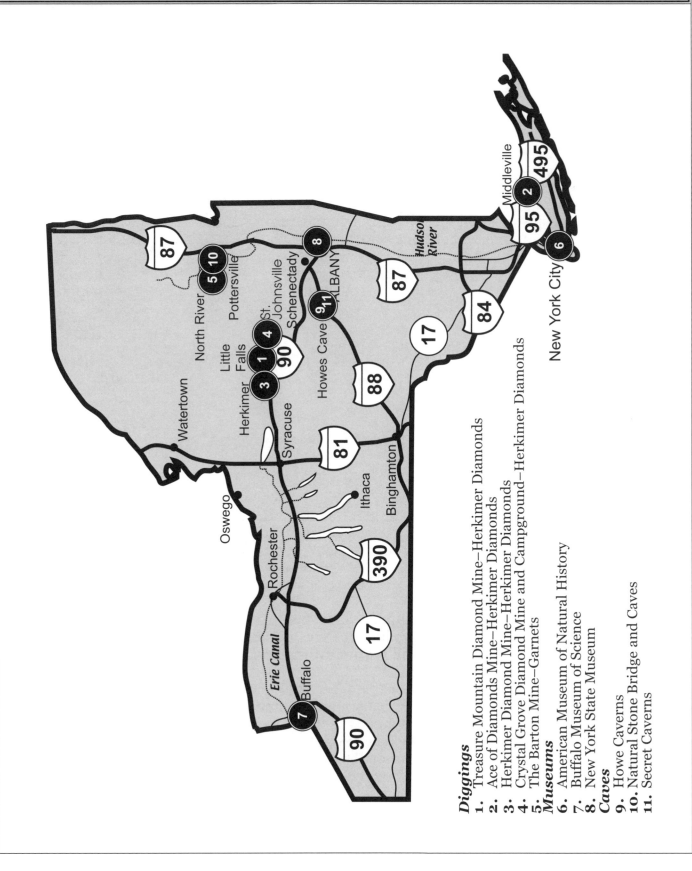

Diggings
1. Treasure Mountain Diamond Mine–Herkimer Diamonds
2. Ace of Diamonds Mine–Herkimer Diamonds
3. Herkimer Diamond Mine–Herkimer Diamonds
4. Crystal Grove Diamond Mine and Campground–Herkimer Diamonds
5. The Barton Mine–Garnets

Museums
6. American Museum of Natural History
7. Buffalo Museum of Science
8. New York State Museum

Caves
9. Howe Caverns
10. Natural Stone Bridge and Caves
11. Secret Caverns

ADDRESS:
Treasure Mountain Diamond Mine
1959 State Route 5S
Little Falls, NY 13365
(315) 823-ROCK
info@treasuremt.com
http://www.treasuremt.com

DIRECTIONS:
The mine is located on State Route 5S in Little Falls, New York. From the New York State Thruway, take Exit 29A. From the toll, travel 0.25 of a mile, then turn left onto Route 169. Travel 0.25 of a mile and turn right on State Route 5S. Drive 1.8 miles to the entrance of the mine.

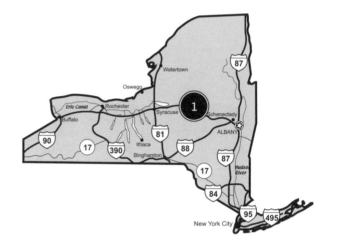

SEASON:
April to November – Weather permitting

HOURS:
8:00 a.m. to dusk

COST:
$10 per day – Adults
$5 per day – Children ages 6 to 12
Free – Children ages 6 and under

WHAT TO BRING:
You can rent the tools you will need or bring your own. You will need a collecting bucket (or two), rock hammer, small sledge, eye protection, gloves and a film container (an X cut in the top to keep your smaller finds safe).

INFORMATION:
This site was recently opened to the public. In its first year, over a thousand visitors came to search for quartz crystals. The site is two hundred seventy acres of open and wooded area. They currently have three staff members and a primitive camping area, with plans to expand and improve operations.

The most unique aspect of this site is the types of "Herkimer" diamonds found here. The locals prefer to call them "Little Falls" diamonds. Specimens at this site are known to form scepters, the rarest and most sought after crystals. Black stemmed scepters topped with quartz crystals are found here and prized. The July/August 2000 issue of Rocks & Minerals features an exquisite example of a scepter crystal on the cover with an article on Little Falls within. In addition to crystals, a layer of shale has been revealed showing numerous fossil specimens. This is a new site and with careful management it will become a favorite in no time.

ADDRESS:
State Highway 28
P.O. Box 505
Middleville, NY 13406
(315) 891-3855 or (315) 891-3896
aceherks@adelphia.net

DIRECTIONS:
The Ace of Diamonds Mine is located on State Highway 28, 8 miles north of Herkimer. Herkimer, New York, is Exit 30 off the New York State Thruway/Interstate Highway 90.

SEASON:
April 1 to November 1

HOURS:
9:00 a.m. to 5:00 p.m.

COST:
$7 – Adults
$2 – Children ages 7 and under

WHAT TO BRING:
Bring standard mining tools, including sledgehammers, safety goggles, wedges and gloves. Tools are also available for rent or for sale at the office. Tools are helpful, but you can find crystals without them. Bring your own food and plenty to drink. Picnic tables are available.

INFORMATION:
You can dig your own Herkimer diamonds here. These doubly-terminated, clear quartz crystals are found in their natural state looking like cut diamonds. The "diamonds" are found inside pockets in rocks, loose on the ground, or by working the cliff wall. Children will have better luck in the tailings piles. A keen eye can find "diamonds" anywhere on the ground, especially after a rain.

Some miners stay for weeks and slowly work at taking down a section of a cliff face using nothing but hand tools (wedges, sledgehammers, crowbars and prybars). They must take the wall down from the top to reach the diamond layer, removing somewhere between ten and thirty feet of stone.

Many day miners find likely rocks on the large tailings piles and spend the day breaking open rocks with hand sledges. Others just turn over rocks to find the diamond beneath. Lastly, miners use screens, garden rakes and the naked eye to search the earth for loose crystals. Crystals can be as large as 4 inches, but you will have to "work the wall" and be very lucky to find one of these. The average crystal is a 1/4- to half-inch long and breathtakingly beautiful.

Camping is available at the mine or at a KOA campground directly across the road.

ADDRESS:
Route 28 North
Herkimer, NY 13350
(315) 891-7355
diamonds@ntcnet.com
http://www.herkimerdiamond.com/

DIRECTIONS:
Take the New York State Thruway (Interstate Highway 90) to Exit 30 to State Highway 28. The mine is 7 miles north of Herkimer on State Highway 28.

SEASON:
April 15 to November 1 – Weather permitting
April to November (camping)

HOURS:
9:00 a.m. to 5:00 p.m.

COST:
$7.50 – Adults (admission includes a visit to the rock museum)
$6.50 – Children ages 5 to 12
Call for group pricing

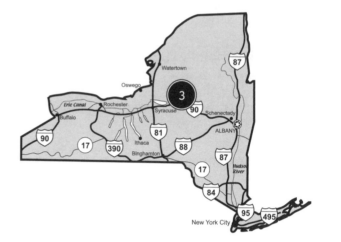

WHAT TO BRING:
Bring standard mining tools, your own food and plenty to drink. A plastic container with a lid, goggles and a large bucket are highly recommended. The mine will let you borrow hammers, chisels and goggles in exchange for your driver's license.

INFORMATION:
These double-terminated, clear, quartz crystals are found in their natural state looking like cut diamonds. Diamond hunting can be hard work. The "diamonds" are found inside air pockets in the local stone. They can be found by turning over rocks in the tailings piles; breaking open likely rocks with hammers; or working the cliff wall. Children will have better luck in the tailings piles. A keen eye can find "diamonds" anywhere on the ground. The people at the mine are friendly and helpful in orienting you. The best candidate rocks are the ones with the dark black pockets visible on the rocks exterior. You can take as many host rocks and diamonds home with you as you find.

There is a playground, picnic tables and a rock and gift shop here. Included with your admission price is a trip to their museum, located above the large rock shop. Across the street from the mine is a KOA campsite on the river. It has a pool, some cabins, bathhouse, playground and store. Fly fishing is another popular hobby in this area.

ADDRESS:
161 County Highway 114
St. Johnsville, NY 13452
(518) 568-2914 or (800) KRY-DIAM
fun@crystalgrove.com
http://www.crystalgrove.com/

DIRECTIONS:
Traveling west on the New York State Thruway (Interstate Highway 90) take Exit 29 (westbound) and turn right onto State Highway 10. Then take State Highway 5 west and proceed to St. Johnsville, New York. Turn right onto Division Street at the town's only traffic light. Drive half a mile and bear right at the fork. Travel 4 miles to the campsite.

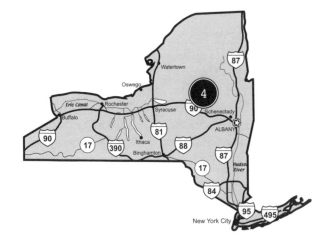

SEASON:
April 15 to October 15

HOURS:
8:00 a.m. to dusk

COST:
$7 – Adults
$5 – Children ages 5 to 14
Free – Children ages 4 and under

WHAT TO BRING:
Standard mining equipment is suggested, along with tweezers, safety goggles and a plastic container with a lid. You may also want a large bucket for carrying home host rocks with good possibilities. Some tools are also available for rent.

INFORMATION:
There are three mines on this property. Herkimer Diamonds can be found in the matrix (rock) or by sifting through the earth. Finding these gems involves breaking up the gray host rock with a rock hammer or sledgehammer. Some rocks are barren, without crystal pockets. Look for stones that have existing, empty and visible pockets on their exterior. The pockets you are looking for are black, dark gray or may glitter with drusy quartz. Pockets lined with white, cubed crystals usually are not hosts to diamonds.

Try to bust open the rock. If it doesn't split with four or five good strikes, it may not have any air pockets, so try again with another likely rock. The best time to go to a Herkimer Diamond Mine is after a good rain. The crystals are exposed and lie glittering on the soil's surface until they dry, becoming invisible once again. Some miners prefer to use a sifting screen to find crystals. Simply shovel small amounts of earth into a screen and see what you find.

Crystal Grove has camping and tent sites by a trout stream, and motor home sites in a wooded area. The camp has a shop selling groceries, rocks and gifts.

ADDRESS:
Garnet Mine Tours
P.O. Box 30
North River, NY 12853
(518) 251-2706
http://garnetminetours.com/mines.html

DIRECTIONS:
Take Interstate Highway 87 to Exit 23. Then follow U.S. Highway 9 through Warrensburg. Go left onto State Highway 28 for approximately 21 miles to North River General Store. Proceed left up the hill (Gore Mountain) on the paved Barton Mines Road for 5 miles to the Barton Mineral Shop on the right-hand side of the road.

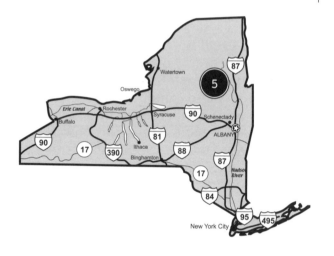

SEASON:
Beginning the last weekend in June to Labor Day –
Open 7 days a week
Labor Day to Columbus Day (weekends only)

HOURS:
9:00 a.m. to 5:00 p.m. – Monday through Saturday
11:00 a.m. to 5:00 p.m. – Sunday

COST:
$9.50 – Adults
$8 – Seniors
$6 – Children
Garnets can be purchased by the pound at a reasonable price.

WHAT TO BRING:
The mine will give you a paper bag for garnets. Bring a plastic bag or small pail to carry your stones. You may find a garden rake and spray bottle helpful in your search. Picnic tables are available, so you may want to bring your own food.

INFORMATION:
This is a tour of a now inactive garnet mine. At one time, this mine produced gem-quality stones. Some garnets were used to make garnet sandpaper and emery boards. Garnets were discovered in the Hudson River, near it's source in the Adirondack Mountains in New York. The garnet trail was followed up the North River to its source on Gore Mountain. Several old quarries remain at this location.

Your visit will begin at the gift shop and headquarters, on top of Gore Mountain. The drive is steep and the road is unpaved. Unless weather conditions are bad, you will have no trouble driving up. From the shop you will follow the tour guide in a car caravan to the quarry. The tour takes place in an open garnet quarry. It includes a brief history of the mine, some geology and thirty minutes of time to gather garnets. The stones are plentiful and easy to find. This is ideal for children.

Museums

Site 6.
American Museum of Natural History
79th Street and Central Park West
New York, NY 10024-5912
(212) 769-5100
http://www.amnh.org/
This museum is colossal with over 90,000 minerals, 20,000 rocks, 4,000 gems (including the Star of India and the Patricia Emerald), a separate hall for meteorites and an entire floor of fossils with over 600 specimens including the tallest free-standing dinosaur (a Barosaurus). Don't forget to visit the new Rose Center for Earth and Space, which includes the Willamette Meteorite and the Hall of Planet Earth dedicated to the processes that formed our planet. This museum is one of the world's best.

Site 7.
Buffalo Museum of Science
1020 Humboldt Parkway
Buffalo, NY 14211
(716) 896-5200
http://www.sciencebuff.org/
This collection contains approximately 100,000 specimens of minerals from around the world with a focus on western New York State, New York and Ontario. The gem displays include objects crafted from minerals by artists.

Site 8.
New York State Museum
Cultural Education Center
Empire State Plaza
Madison Avenue
Albany, NY 12234
(518) 474-5877
http://www.nysm.nysed.gov/
Located in the Cultural Education Center, this museum houses exhibits of relevance to the state. Among them is the Minerals of New York exhibit in Adirondack Hall. Minerals of surprising variety and beauty are displayed in natural and gem form. Exhibits change periodically to allow the public to see more of the vast mineral holdings, including an impressive display of the State gemstone, the garnet.

Caves

Site 9.
Howe Caverns,
255 Discovery Drive
Howes Cave, NY 12092
(518) 296-8900
http://www.howecaverns.com
This cave was known by Native Americans and a few early settlers, but the knowledge seems to have been lost until the 1800s when a farmer, named Lester Howe, noticed a strange "blowing rock" where a cool breeze always emanated. He discovered the cave and began giving tours by lantern light. Now visitors use an elevator and electric lights to tour the cave's wonders, which include a boat ride, down an underground river. This cave is a popular spot for weddings.

Site 10.
Natural Stone Bridge and Caves
535 Stone Bridge Road
Pottersville, NY 12860
(518) 494-2283
http://www.stonebridgeandcaves.com
This cave was described in Morse's Geography, written in 1790. Modern visitors use a well marked trail to discover the unique beauty of Stone Bridge and Caves on a self-guided tour.

Site 11.
Secret Caverns
Caverns Road
Howes Cave, NY 12092
(518) 296-8558
http://www.secretcaverns.com/
According to the owners, two unlucky cows strayed from their pasture in the year 1928 in search of a cool breeze coming from a hole in the earth. The cows fell eighty-five feet and the farmer who went searching for his errant livestock soon discovered the cave's natural entrance. Further exploration revealed soaring chambers, spectacular limestone formations and a 100-foot waterfall. Visitors descend into the cavern by staircase and not by plummeting 85 feet. Don't worry - the cows are gone.

Home of the first gold discovered in the United States and the largest cut emerald on the continent.

America's first gold rush started in North Carolina. In 1799 a boy named Conrad Reed found a shiny golden stone on his father's farm in Cabarrus County and brought it home. The family failed to recognize the heavy lump as a gold nugget, and used the metal as a doorstop for many years and finally sold the ore to an unscrupulous jeweler who paid $3.50 for the 17-pound nugget worth thousands of dollars. When Reed learned the nugget was gold, he began prospecting. In 1803 a slave recovered a gold nugget weighing 28 pounds. This discovery caused much regional excitement.

By 1820 shafts were sunk into gold bearing quartz veins. Gold fever hit and thousands rushed to North Carolina in search of wealth. By 1830 fifty-six gold mines operated in this region and North Carolina became America's first golden state. But North Carolina is known for another first.

Area miners were reluctant to turn over their hard earned gold dust for the government's paper script. They wanted coins - gold coins. The constitution prohibits any state from minting coins, but it does not prevent individuals from doing so. A German metal smith and jeweler named Christopher Bechtler began assaying and minting gold. He charged four percent for his work and was known for reliability. His coins were the first gold currency minted on United States soil. Gold brought miners to North Carolina and prospectors are still there. Many gold mining areas are open to the public, and hunting for gold is a popular hobby for locals. Prospectors do not have to head to California to learn the ins and outs of gold mining. All you need is in North Carolina. But if you'd rather find precious gems, look no further.

Emerald is North Carolina's official state stone with good reason. The area near Hiddenite in Alexander County is known to yield fine quality emeralds. In 1969 a 1,438-carat stone was unearthed at the Rist Mill at Hiddenite. The resulting 12.14-carat emerald-cut stone was dubbed "the Carolina Emerald" and valued at $100,000 dollars at the time. This is the largest fine emerald ever cut on the continent. This distinction is impressive enough, but North Carolina has much more.

Star rubies, pink sapphires and blue star sapphires are all found in North Carolina. Transparent pink sapphire crystals, shaped like thin pencils, are found at the Mason Mine. These little beauties are recovered from the clay and kaolin by washing and scrubbing gems from clay using screening boxes. For rubies try the Old Cardinal Mine where visitors sift through the gravel for star rubies suitable for cabochon cutting. The Old Pressley Mine yields star sapphires ranging from cornflower blue to gray. Those in search of emeralds can try their luck at the Emerald Hollow Mine or Vein Mountain Gold & Gems. Be advised that emeralds are very rare and very valuable. These stones are not found in great quantity, but other mining products are.

North Carolina leads the country in the production of feldspar used to make components in television picture tubes, computer components, bath tiles, abrasives, ceramics, plumbing fixtures, specialty glass, cosmetics, asbestos replacement, filters and shingles. Additional mined resources include kaolin clay for the manufacture of dinnerware and porcelain; spodumene for lithium used in spacecraft, aircraft, batteries, grease, photography and paint; mica for use in cosmetics and as an insulator; olivine for lining bake furnaces and kilns; talc and pyrophyllite for products such as soaps, bleaching powder, talcum powder, electrical insulators; and clay for brick manufacture. Phosphate is mined for plant food, fertilizer, pesticides and photography, while granite is quarried for building stone.

North Carolina offers rockhounds and prospectors a great variety of adventures. No other state has this variety of gemstones available for visitors to mine. When one considers the gold, gemstones and variety of mineral wealth, North Carolina is truly blessed in the riches of the earth.

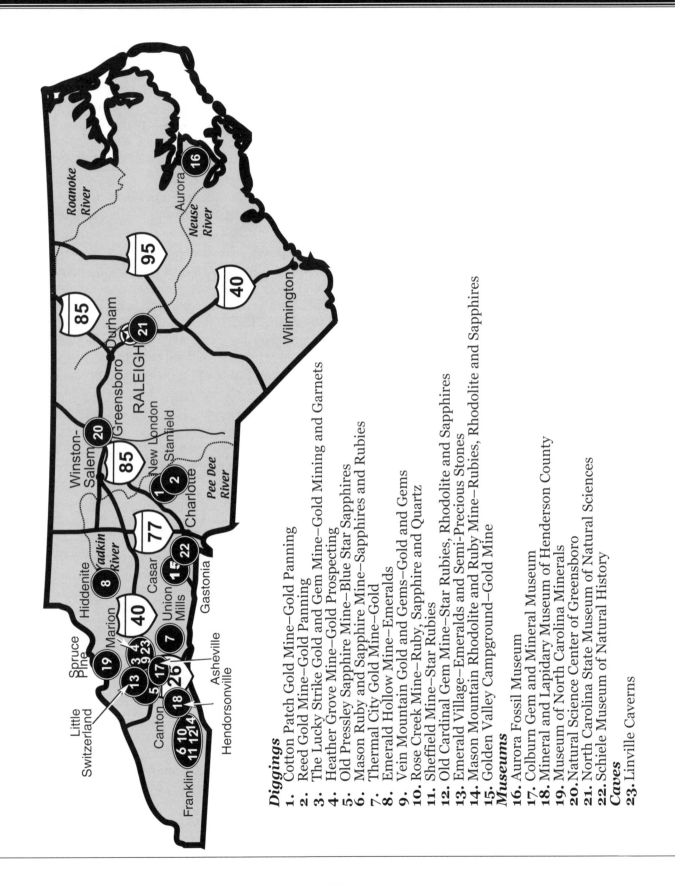

Diggings
1. Cotton Patch Gold Mine–Gold Panning
2. Reed Gold Mine–Gold Panning
3. The Lucky Strike Gold and Gem Mine–Gold Mining and Garnets
4. Heather Grove Mine–Gold Prospecting
5. Old Pressley Sapphire Mine–Blue Star Sapphires
6. Mason Ruby and Sapphire Mine–Sapphires and Rubies
7. Thermal City Gold Mine–Gold
8. Emerald Hollow Mine–Emeralds
9. Vein Mountain Gold and Gems–Gold and Gems
10. Rose Creek Mine–Ruby, Sapphire and Quartz
11. Sheffield Mine–Star Rubies
12. Old Cardinal Gem Mine–Star Rubies, Rhodolite and Sapphires
13. Emerald Village–Emeralds and Semi-Precious Stones
14. Mason Mountain Rhodolite and Ruby Mine–Rubies, Rhodolite and Sapphires
15. Golden Valley Campground–Gold Mine

Museums
16. Aurora Fossil Museum
17. Colburn Gem and Mineral Museum
18. Mineral and Lapidary Museum of Henderson County
19. Museum of North Carolina Minerals
20. Natural Science Center of Greensboro
21. North Carolina State Museum of Natural Sciences
22. Schiele Museum of Natural History

Caves
23. Linville Caverns

ADDRESS:
41697 Gurley Road
New London, NC 28127
(704) 463-5797
cottonpatchgoldmine@vnet.net
http://www.cottonpatchgoldmine.com/default.htm

DIRECTIONS:
From Richfield, North Carolina, take U.S. Highway 52 to State Highway 740. Travel through New London and turn left on Gurley Road to the mine.

SEASON:
March 1 to December 15

HOURS:
9:00 a.m. to 5:00 p.m. – Wed. to Sun.

COST:
$8 per person – Panning fee for five, two-gallon buckets of dirt
$4 – Children ages 12 and under
$16 – Sluicing fee for three, five-gallon buckets of dirt

WHAT TO BRING:
The mine will supply you with all the materials you need. You may bring your own standard mining tools. Motorized mining tools are not permitted.

INFORMATION:
The Cotton Patch Gold Mine is a place where you can learn to pan and sluice for gold. The mine offers instruction to get you started. You may buy dirt by the bucket or by the front-end load. For sluicing, the costs range from $35 for a quarter-load to $95 for a full load. This material is brought from a gold-producing area of the property. Panning troughs are provided to help you clean your concentrates. The stream has notched logs to ease the setup of your sluice box.

You will find picnic areas here and a general store that sells prospecting supplies, food and beverages. Camping with and without electric hookups is available. This is a fun, educational adventure for the entire family.

On the last Sunday of October, the Cotton Patch Gold Mine hosts a gold panning championship.

ADDRESS:
9621 Reed Mine Road
Stanfield, NC 28163
(704) 721-GOLD
reedmine@ctc.net
http://www.he.net/~brumley/tour/daytrips/reedgoldmine.htm

DIRECTIONS:
From Interstate Highway 85, exit at Salisbury, North Carolina. Travel south on U.S. Highway 601 through the town of Concord to State Highway 200. Bear left onto State Highway 200 and drive 4 miles to Reed Mine Road. Turn left at the Reed Gold Mine sign into the mine.

SEASON:
Year round

HOURS:
April through October: 9:00 a.m. to 5:00 p.m. – Monday to Saturday, 1:00 p.m. to 5:00 p.m. – Sunday
November to March: Closed Monday, 10:00 a.m. to 4:00 p.m. – Tuesday to Saturday, 1:00 p.m. to 4:00 p.m. – Sunday

COST:
Free – Tours of mine and stamp mill
$2 – Fee for gold panning

WHAT TO BRING:
Mining supplies are provided by the mine.

INFORMATION:
The Reed Gold Mine reported the first gold find in the United States. As the story goes, in 1799, Reed's son, Conrad, found a large yellow rock in the Little Meadow Creek while playing hooky from church. A silversmith in Concord was unable to identify the rock, but measured it's weight at seventeen pounds. The Reeds used it as a door stop until a jeweler bought it in 1802, for $3.50, less than one-tenth of its value. Later Reed discovered his mistake and "convinced" the jeweler to compensate him.

Mining later began by Reed and others and a twenty-eight-pound nugget was found. A total of $100,000 worth of gold was unearthed by 1824. The Reed Gold Mine has an underground tour, stamp mill for crushing rock, and a panning area. This area is open for panning only. The Reed Gold Mine is worth a visit for historical interest, but don't expect to find a twenty-pound nugget, because you are hundred years too late.

ADDRESS:
The Lucky Strike Gold & Gem Mine
251 Lucky Strike Drive
Highway 221 South
Marion, NC 28752
(828) 738-4893
theluckystrike@icu2.net
http://www.luckystrikegoldandgem.com/

Direction:
From Marion, North Carolina, take Interstate Highway 40 to U.S. Highway 221. Travel 5 miles south on U.S. Highway 221 and turn left onto Polly Spout Road. Follow the road 0.3 mile to the mine entrance.

SEASON:
Open all year round – Weather permitting

HOURS:
8:00 a.m. to 6:00 p.m. daily

COST:
Call for prices

WHAT TO BRING:
Some supplies are available for rent including pans, dredges, highbankers and sluice boxes. You may use the troughs, poop tube and spiral wheel for concentrating your material for no charge. Bring standard gold mining equipment.

INFORMATION:
The Lucky Strike Gold Mine has all you need for a gold mining adventure. You will find a roofed area with several flumes. Here you can pan for some gemstones or get a gold panning lesson. This is also where the gold miners will gather in the evening to separate their concentrates. Most miners who work this river will find some color. Many nuggets have been recovered here, as well as fine gold and flakes. The gold is often found in conjunction with the host quartz.

If you plan to camp, you can sluice or pan all day in the river for no additional charge. Lessons on dredging are given, and you will learn all you need to know. You may also want to try high banking. The mine holds common digs throughout the year. This is an opportunity to join forces with others and have a weekend of gold mining. All the gold you find in the common digs is divided equally among the participating miners. Call the mine to find out more details.

Campsites with hookups are available right by the river. Restroom and shower facilities are also available. The Lucky Strike Gold Mine's "Miner Diner" has reasonably priced food for those too tuckered to cook. The owners are friendly and helpful.

ADDRESS:
43 Polly Spout Road
Marion, NC 28752
(828) 738-3573
mikeandbrooke@heathergrovegold.com
http://www.heathergrovegold.com/index.shtml

DIRECTIONS:
From Interstate Highway 40 at Marion, North Carolina, take U.S. Highway 221 south to Polly Spout Road. Look for the mine immediately on your left.

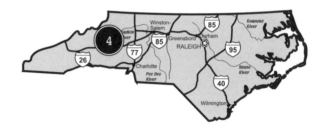

SEASON:
Open all year round (heated, indoor panning area available)

HOURS:
Daylight

COST:
$10 – Fee for a three-gallon gem buckets
$10 – Fee for a three-gallon gold buckets
$6 per day – Fee for sluicing and panning
$10 to $15 – Fee for dredging
$25 – Fee for high banking
$6 – Fee for a metal detecting or dowsing

WHAT TO BRING:
Equipment is available to rent or bring your own gold mining equipment.

INFORMATION:
Opened in 1996, this nine-acre property is situated on a known gold-bearing stream beneath Vein Mountain. The mine offers a variety of recreational mining opportunities. You can sift through buckets for gems or pan out buckets of dirt. Or you can try your luck in the stream with a pan, sluice, highbanker or dredge.

Heather Grove is a family-oriented operation. The mine has a gift shop with supplies, food and gifts for sale. RV and tent camping sites are available with or without hookups. There are also several new cabins with shower facilities for rent. One cabin is handicap accessible. There is a bathhouse for campers. Pets are permitted, but must be kept on a leash, so they do not disturb the owner's pet chickens and ducks.

ADDRESS:
240 Pressley Mine Road
Canton, NC 28716
(828) 648-6320 or (877) 903-4754
srichardson3@ec.rr.com
htttp://www.oldpressleysapphiremine.com

DIRECTIONS:
Take Exit 31 off Interstate Highway 40 and go left to Canton, North Carolina. Drive past the paper factory and up the hill (away from the business district). Turn left onto North Hominy Road and follow the signs. Proceed left onto a dirt road. The Old Pressley Mine is the second mine on this dirt road.

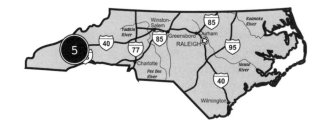

SEASON:
Open all year round
April 1 to October 31 (flumes open)

HOURS:
9:00 a.m. to 6:00 p.m.– Open 6 days a week
April to October
10:00 a.m. to 5:00 p.m. – Open 6 days a week
November to March

COST:
$5 – General Admission
Group rates available
$1 – a bucket to process dirt

WHAT TO BRING:
The mine supplies shovels, buckets, sifting screens and sluices. Bring your lunch or cooler. Chips and soda are available.

INFORMATION:
This is the home of the "Southern Star," the world's largest blue sapphire at 1,035 carats. The mine is set in a beautiful spot in Canton, North Carolina. The folks at the mine are friendly and helpful. They will answer all questions and offer tips on how to search. The sapphires that are found here are large in size. You will find blue star sapphires, but also gray, blue-gray, sky blue, corn silk blue and occasionally a pink sapphire.

Save all heavy stones that are covered with white feldspar and mica and have the experts at the mine check them for you. You may have missed a sapphire. A small rock shop is found in the office. Restrooms are available, but bring your own fresh water as the facility uses pond water.

ADDRESS:
7091 Upper Burningtown Road
Franklin, NC 28734
(828) 369-9199

DIRECTIONS:
Franklin, North Carolina is located approximately 25 miles south of Great Smoky Mountains National Park, and 60 miles southwest of Asheville, North Carolina. From Franklin, take State Highway 28 north toward the airport and look for the Mason sign on the left. Turn left onto State Road 1392/Upper Burningtown Road. It is 8.5 miles to the mine.

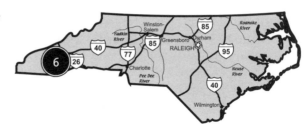

SEASON:
April 1 to October 31

HOURS:
9:00 a.m. to 5:00 p.m. daily

COST:
$10 – Adults
$7 – Children ages 8 to 10
(fee includes digging all day and all equipment)

WHAT TO BRING:
All equipment is supplied, including shovels, buckets, sorting screen boxes, and even a film container to hold your loot. Bring your lunch. A picnic area is available. Some miners wear rubber gloves because the water in the sluice is cold.

INFORMATION:
Many of the mines in this area are seeded. This means they add stones to the dirt. The stones are from all over the world and are usually of poor quality. Most unseeded mines are a bit off the beaten track. This area has over twenty mines. Check with the Franklin Chamber of Commerce for details. Indigenous stones to Franklin County include sapphires and rubies.

The Mason Mine is not seeded and the digging is easy. You can have the stones cut in town at Ruby City and search through their shop at 130 East Main Street in Franklin (828) 524-3967. The workers at the mine will explain the sifting of materials and how to spot the cylindrical sapphires. These stones are a clear, light pink when cut.

ADDRESS:
Thermal City Gold Mine
5420 U.S. 221 North Highway
Union Mills, NC 28167
(828) 286-3016
jnanney@blueridge.net
http://www.huntforgold.com/thermalcity3.htm

DIRECTIONS:
The Thermal City Gold Mine is located off Interstate Highway 40 between Asheville and Hickory, just south of Marion, North Carolina. Take Interstate Highway 40 to Exit 85. Then take U.S. Highway 221. Drive 5 miles south on Highway 221. Turn left onto Polly Spout Road. Drive approximately 5 miles down this road past Heather Grove, Lucky Strike and Vein Mountain Gold & Gem Mines.

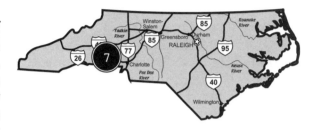

SEASON:
March 1 to November 30

HOURS:
8:30 a.m. to 5:00 p.m.

COST:
$5 – Fee per person per day for panning, detecting and sluicing
$15 – Fee for camping with power hookup ($10 – Fee for primitive camping)
Dredging and Highbanking are also available. Call for prices and details.

WHAT TO BRING:
Thermal City features gold panning and a washing area. You may use their equipment or bring your own.

INFORMATION:
Thermal City opened in 1992. They have one mile of a known gold-bearing river (the second broad) and thirty acres of land. The river provides a great work area. You can pan there or try material brought from the river with a backhoe. The owners tell us that the gravel is not enriched.

North Carolina was the site of the first American gold rush in the 1830's. This site was mined at that time. Today you can spend a day or two mining where it all began. Thermal city also holds four common digs each year between May and September. The operation uses a backhoe fed wash plant to extract gold. Common digs cost $100 per miner for two days. Miners take home their share of gold and black sands.

ADDRESS:
Emerald Hollow Mine
P.O. Box 276
Hiddenite, NC 28636
(828)632-3394

DIRECTIONS:
Hiddenite, North Carolina is located 60 miles north of Charlotte and 20 miles north of Hickory. To get to the Emerald Hollow Mine, take Interstate Highway 40 and exit at North Carolina Route 90. Travel west 12 miles to the town of Hiddenite. Turn right at Sulphur Springs Road (State Route 1001). Drive half a mile and watch for the Emerald Hollow sign on the right. Drive 0.125 of a mile on the sand/dirt road to the parking area at the mine.

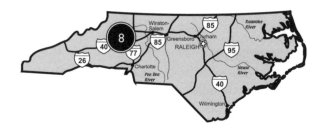

SEASON:
Open all year round

HOURS:
8:30 a.m. to sunset

COST:
$5 – Fee to pan through unaltered pre-dug dirt
The also have buckets of soil seeded with semi-precious stones–a popular choice for children. Prices vary for digging your own pit at the mine and sluicing or panning in the creek.

WHAT TO BRING:
Visitors use fine mesh screens to sift through the red dirt to recover the gems and minerals hidden there. If you are planning to dig, bring a shovel, buckets and work gloves. Equipment is available to rent for a nominal fee.

INFORMATION:
The town is named after W.E. Hidden who discovered the clear green crystals similar to emeralds, but different in chemical composition, a variety of spodumene. Mr. Hidden was searching for platinum for inventor Thomas Edison when he discovered the rare gem, found only in the few acres around the town.

Visitors may also unearth rubies, sapphires, aquamarines, kunzite, tourmaline, amethyst and garnets, among others.

Emeralds and hiddenite are extremely uncommon gems and finding them is rare. Emerald Hollow is the only mine in the county open to the public.

ADDRESS:
1694 Polly Spout Road
Marion, NC 28752
(828)738-9544
gemmine@wnclink.com
http://home.wnclink.com/gemmine/

DIRECTIONS:
From Marion, North Carolina, take Interstate Highway 40 to U.S. Highway 221. Turn left onto Polly Spout Road. Travel south past the Heather Grove Mine and the Lucky Strike Mine approximately 3.5 miles from the start of Polly Spout Road.

SEASON:
Open April 1 to November 5

HOURS:
Daylight

COST:
$6 to $10 – Fee for enriched material
$6 to $10 – Fee for hand sluicing
$12 to $20 – Fee for highbanking
$12 to $20 – Fee for dredging
Camping is available. A new bathhouse with handicap facilities and private showers, picnic tables and a fire pit are all available. The hookups are thirty/fifty amps.

WHAT TO BRING:
Some rentals are available. You are welcome to bring your own equipment to sluice, pan, highbank or dredge.

INFORMATION:
The owner, Don Davidson, has thirty acres of land on the second broad river in a historic gold bearing area. With Don's permission, customers can dig from the riverbank or the hillside. This area is known to produce feldspar, garnets, black tourmaline, light green tourmaline, green aquamarine, quartz, mica and gold. All these markers are indicators for emeralds. Rockhounds are welcome to try their luck. Don is actively seeking emeralds and has found all the indications that they are here, including a pegmatite vein.

ADDRESS:
Rose Creek Mine
115 Terrace Ridge Drive
Franklin, NC 28734
(828) 349-3774
rocks@smet.net
http://www.shipwreckcoins.com/rose.htm

DIRECTIONS:
Located on Highway 28, 4.5 miles north of Franklin, North Carolina. Turn left before the bridge onto Bennett Road and drive 0.125 of a mile, then turn right on Rose Creek Road. Drive up the hill, turn left on Lyle Downs Road, and left at the sign. When you see the big red barn you are there.

SEASON:
April to October

HOURS:
9:00 a.m. to 5:00 p.m. – Open 7 days a week

COST:
$5* – Adult mining fee
$3* – Child mining fee
(*includes one free bucket, refills are $2 per bucket)
$3 per bag – Fee for gem dirt (native & enriched earth)
$10 per person – Fee for tunnel digging (native stones, non-enriched)
$3 per pound of trout – Fee for trout fishing (includes bait and tackle)
$15 per night – Camping fee

WHAT TO BRING:
All equipment is provided.

INFORMATION:
Rose Creek has much to offer. Gem bags with native and enriched specimens will keep children happy and entertained. Assistance is offered to beginners. Miners can have at the virgin earth and find native gems. If you are lucky you may find a ruby or sapphire. They are known to be found in this area.

If you tire of gem hunting you can try a hand at fishing. The mine provides the bait and tackle. Snacks can be purchased on site and restrooms are available.

ADDRESS:
Sheffield Mine
385 Sheffield Farms Road
Franklin, NC 28734
(828) 369-8383
ruby@sheffieldmine.com
http://www.sheffieldmine.com

SEASON:
April 1 to October 31

HOURS:
10:00 a.m. to 5:00 p.m.– Open 7 days a week,
rain or shine
Arrive before 3:30 p.m. to have time to look
for rubies and sapphires

COST:
$8 – Adult fee (includes two buckets)
$.75 – Each additional bucket

WHAT TO BRING:
The mine provides you with all the equipment you will need to mine for gems. Wear old clothes!
Sheffield Mine knows that miners get a bit wet and dirty. They even have a formula to get that
red mud out of your duds. Use your regular detergent, plus Shout™ and half-cup of Oxi-Clean™
or Quick 'N Brite™

INFORMATION:
On July 17, 2000 a visitor to the Sheffield Mine found a star ruby weighing 320 carats!
Rubies over 30 carats are called "Honkers" and are the best candidates to have cut. Native
rubies are of the star variety with six points. The only other place they are found is Burma.

The owners suggest you have at least two hours to search. The process is slow, relaxing and
very addictive. Searching the native dirt takes about four times longer than going through the
salted material. The dirt and gems in the native soil have been together for millions of years and
it's hard to break up a relationship like that!

Please arrive before 3:30 p.m., so you'll have time to finish your buckets. No native dirt is
sold after 4:00 p.m. If you prefer fast results try the "Rainbow" buckets, which are enriched with
semi-precious stones from all over. One in ten of these buckets contains an arrowhead. The
mine sells snacks & cold drinks, but recommends you bring a lunch. There is also a gift shop
and restrooms on site.

ADDRESS:
Old Cardinal Gem Mine
71 Rockhaven Drive
Franklin, NC 28734
(828) 369-7534
http://www.oldcardinalgemmine.com/

DIRECTIONS:
From Franklin, North Carolina take Highway 28 north approximately 5.5 miles. Watch for the sign to Mason Branch Road. Turn right and drive a few 100 yards to the mine entrance.

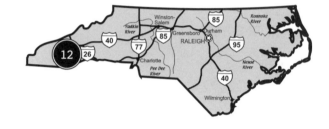

SEASON:
March 1 to October 31

HOURS:
8:00 a.m. to 5:00 p.m. – Open 7 days per week

COST:
$5 – Adults
$3 – Children
$2 per bag – Fee for dirt
$10 – Fee for six bags

WHAT TO BRING:
Wear old clothes and shoes and bring a lunch if you plan to linger. Equipment for sifting and washing your dirt is provided by the mine. If it's cold or you have a manicure, wear rubber gloves. A ziplock bag is handy to store your gems.

INFORMATION:
Among the native stones found here is pink rhodolite. This is a rare and uniquely colored rose garnet found only in this valley. The mineral gets its name from the Greek words for rose and stone (rose-stone). Also found here are rubies, sapphires and moonstone.

The flumes are in the shade so you can perch and sift out of the heat. There is a picnic area on site. Cold drinks and snacks are available along with restrooms.

The mine offers help to beginners, and has enriched soil for children or adults who want guaranteed results.

ADDRESS:
Emerald Village
P.O. Box 98
Little Switzerland, NC 28749
(828) 765-6463
mine@wnclink.com
http://www.emeraldvillage.com/default.htm

DIRECTIONS:
From Little Switzerland take Chestnut Grove Church Road west about 1 mile to McKinney Mine Road. Drive approximately 2 miles to the mine.

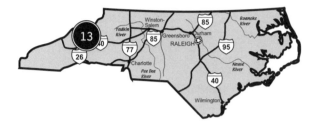

SEASON:
April 1 to Labor Day

HOURS:
9:00 a.m. to 5:00 p.m. – April 1 to Memorial Day
9:00 a.m. to 6:00 p.m. – Memorial Day to Labor Day

COST:
$5 – Adults
$4.50 – Seniors
$4 – School-aged children
Free – Preschool children

WHAT TO BRING:
You should bring a picnic lunch and a camera.

INFORMATION:
High in the Blue Ridge Mountains, near Little Switzerland is a group of mining attractions called Emerald Village. This area has a rich mining history and boasts forty-five different native rocks and minerals, including emeralds, of course.

From the Visitor's Center you may take a self-guided tour of Bon Ami Mine, which showcases mining equipment. The Gem Mine provides a place to shift through mined earth salted with additional material so you are guaranteed to find semiprecious stones. There are gem cutters on site to turn your finds into jewelry. In the Company Store you will see a miniature mining town reminiscent of life in the 1920s. The Gallery of Minerals houses rocks and minerals from around the world. A collection of ultraviolet minerals is also on display. These rocks fluoresce under a black light to show astonishing neon colors.

ADDRESS:
Mason Mountain Rhodolite and Ruby Mine
5315 Bryson City Road
Franklin, NC 28734
(828) 524-4570
http://www.keysiteservices.com/members/tjrocks/masonmtnmine.html

DIRECTIONS:
Located on Highway 28 (Bryson City Road).
Drive 5.5 miles north on Highway 28.

SEASON:
May 1 to November 1
April and November (weekends only)

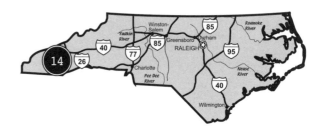

HOURS:
8:30 a.m. to 5:30 p.m. – Open 7 days a week

COST:
$5 – Adults
$3 – Children ages under 12
$2 – Buckets of dirt
$10 to $20 – Special gem bag
$25 to $300 – Super buckets

WHAT TO BRING:
The mine will furnish all mining equipment. Bring a lunch, rubber gloves and something in which to stash your finds.

INFORMATION:
    This mine hasn't been in business for twenty years for nothing. They offer it all, soup to nuts. You can choose native buckets or enriched material. It is possible to dig a native ruby, rhodolite or sapphire from the soil; choose a setting from the gift shop in sterling or 14 karat gold; and have your stone cut and set at the mine. There are experienced gem cutters on site.

    The flume area is covered to protect you from sun and rain. Snacks, sandwiches and soda are available. The mine also has a picnic area and restrooms. There is a gift shop and a gem shop on site.

ADDRESS:
680 NC Highway 226 South
Casar, NC 28020
(704) 538-8047

DIRECTIONS:
Golden Valley is located on State Route 226 between Marion, North Carolina and Shelby, North Carolina. From Interstate Highway 40, take Exit 86 and drive 22 miles. Golden Valley Campground is on the left hand side.

SEASON:
April 1 to October 31

HOURS:
7:00 a.m. to 7:00 p.m.

COST:
$5 – Sluicing fee per day $30 – Sluicing fee per week
$5 – Highbanking fee per day $30 – Highbanking fee per week
$5 – Fee for two-inch dredge per day $30 – Fee for two-inch dredge per day*
$8 – Fee for three-inch dredge per day $50 – Fee for three-inch dredge per day*
$10 – Fee for four-inch dredge per day $65 – Fee for four-inch dredge per day*
$5-$10 – Fee for gem buckets
$5-$15 – Fee for camping per day
* rates are for campers. If you are not camping, rates are higher.

WHAT TO BRING:
Equipment rentals are available or bring your own.

INFORMATION:
Golden Valley Campground is located on the gold-bearing Little Broad River on sixty-three acres. This peaceful valley was once the sight of the first American gold rush. This site opened in the Spring of 2001 and provides visitors a place to try some gold prospecting. You can buy a bucket and learn to gold pan or jump in the river with a four-inch dredge. According to the manager, a white quartz arrowhead and emerald were recently found while creating an access road to the river.

Museums

Site 16.
Aurora Fossil Museum
400 Main Street
Aurora, NC 27806-0352
(252) 322-4238
http://pamlico.com/aurora/fossils/
This museum centers on the geologic forces that created the Coastal Plains and includes two rooms of fossilized bone, teeth, shell and coral.

Site 17.
Colburn Gem & Mineral Museum
2 South Pack Square
Asheville, NC 28802
(828) 254-7162
http://www.colburnmuseum.org/
A collection of over 4,500 rocks and minerals specimens from North Carolina and around the world are included with fluorescent minerals, crystals, fossil and 1,000 cut gemstones.

Site 18.
Mineral & Lapidary Museum of Henderson County
400 North Main Street
Hendersonville, NC 28792
(828) 698-1977
http://www.mineralmuseum.org/
Exhibits of minerals, fossils and Native American artifacts of North Carolina include two large examples of petrified wood, and many colorful geodes. In addition there is a display dedicated to the Cullinan, a 3,106.75 carat South African white diamond found in 1905. This exhibit includes replicas in cubic zirconium of all nine major stones cut from this amazing gemstone.

Site 19.
Museum of North Carolina Minerals
79 Parkway Road
Spruce Pine, NC 28777
(828) 765-2761
http://www.mitchell-county.com/festival/museum.html
Located at Milepost 331 on the Blue Ridge Parkway, this museum exhibits minerals found in the Spruce Pine Mining District and offers demonstrations on cutting, polishing and shaping stone.

Site 20.
Natural Science Center of Greensboro
4301 Lawndale Drive
Greensboro, NC 27455
(336) 288-3769
http://www.natsci.org
This museum includes the Gem and Mineral Gallery, an amethyst crystal cluster several feet long, and dinosaur fossils.

Site 21.
North Carolina State Museum of Natural Sciences
11 West Jones Street
Raleigh, NC 27601-1029
(919) 733-7450 or (877) 4NATSCI
http://www.naturalsciences.org/
The geology collection has 7,500 specimens largely from North Carolina, including meteorites, historic gold samples, vein quartz, and specimens from the state's historic corundum and monazite mining operations.

Site 22.
Schiele Museum of Natural History
1500 East Garrison Boulevard
Gastonia, NC 28054
(704) 866-6900
http://www.schielemuseum.org/
The Schiele Museum includes an exhibit called, "The Hall of Earth and Man," which includes fossils, dioramas, models and artifacts used to illustrate the evolution of life on Earth and contains North Carolina's largest gem & mineral display.

CAVES

Site 23.
Linville Caverns
Linville Falls
P.O. Box 567
Marion, NC 28752
(828) 756-4171 or (800) 419-0540
http://www.linvillecaverns.com/
The mysterious fish swimming out of Humpback Mountain led to the discovery of this cavern system in the 1800s. The series of rooms and passages containing elegant shapes and textures were created by water. Tours include information about geology, legends and stories about this special cave.

A most valuable mud

If you had visited Ohio 260 million years ago, you would have needed a boat. This state was once a tropical ocean filled with plants, mollusks, coral and fish. Over time the ocean receded and left a swamp. Fossil remains from the ocean were covered with a thick vegetative layer, which gradually transformed into peat, then lignite and finally soft coal. A glacier came and went leaving a thick deposit of mud over much of the land.

Proof of the ocean's existence is the large coral reef remnant at the Falls of Ohio State Park just over the state line in Indiana. Visiting a coral reef usually requires a mask and fins, but here guests stay above sea level at the interpretive center before exploring this fascinating fossilized reef. Small marine fossils are plentiful here and around Griggs Reservoir and Alum Creek. Occasionally a vertebrate fossil such as a tooth or fish skeleton is found. Another interesting fossil left from the sea are cricoids. These ocean animals resemble plants, but are closest in relation to the starfish. Anchored to the ocean floor by a root system, they grow on long stalks like a plant stem constructed of button shaped skeletal discs. When the animal dies these disks remain behind, sometimes fossilizing into odd little disks. Local Native Americans prized them. The fossils were threaded into necklaces and used on ornamental objects and therefore became known as Indian Beads.

The Adena Indians used local flint to make tools and weapons. Flint is one of the few natural resources in Ohio now in short supply. Other resources are plentiful including coal, fire clay, salt and dimension stone.

Fire clay was first used by the Ohio Indians to make pottery, and later by European settlers to make brick. Soon the clay was used to make pottery on potter's wheels. Ohio continues to make use of this resource with industries dedicated to making brick, brick face, pottery and tile. This fire clay withstands high temperatures without warping making it ideal for fashioning building materials and ceramics. These industries comprise millions of dollars in the state's economy, making it a very valuable mud.

Shale occurs in abundance in Ohio, covering nearly two-thirds of the state, and is quarried for use in building.

Coal was first noted by frontiersmen in 1748, who noticed a deposit on fire. Coal mining began in the 1800s with immigrant Scots, Welsh and English men providing the labor force. In the 1830s coal replaced wood for fuel in industry, railroad and homes making mining more lucrative. Mules, ponies, goats and even dogs were enlisted to haul coal from the mines. Today few underground coal mines exist. Surface mining is more profitable, so the number of miners in the state had dwindled from 50,000 at the beginning of the twentieth century to less the 4,000 at the end.

In 2001 an Ohio State Geologist working in a coal mine made an interesting discovery. They recovered a fossil of a 300 million year old cockroach that is nearly 4 inches long. The roach rambled Ohio long before the dinosaurs. Plant fossils are more commonly found, but the big bug caused quite a stir. You never know what you'll find when digging in the earth.

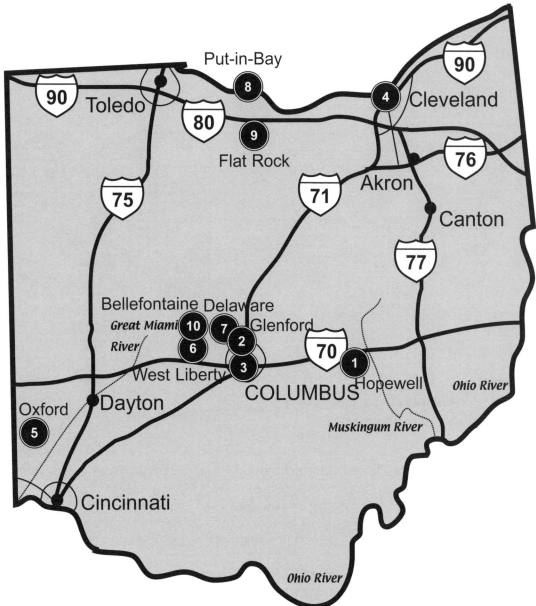

Diggings
1. Hidden Springs Ranch–Flint
Museums
2. Flint Ridge Museum
3. Orton Geological Museum
4. The Cleveland Museum of Natural History
5. Karl E. Limper Geologic Museum
Caves
6. Ohio Caverns
7. Olentangy Indian Cavern
8. Perry's Cave
9. Seneca Caverns
10. Zane Shawnee Caverns

ADDRESS:
Hidden Springs Ranch
9305 Hidden Springs Road
Hopewell, OH 43746
(614) 787-2060
genie@infinet.com

DIRECTIONS:
Call or e-mail for directions.

SEASON:
Open year round

HOURS:
Open by appointment only
8:00 a.m. to 6:00 p.m. each day

COST:
Adults - $5.00
Children $3.00
$3.00 fee for each five gallon bucket of flint.

WHAT TO BRING:
Shovel, pry bar, rock hammer, buckets, water
and a bristle brush. Gloves are a must.

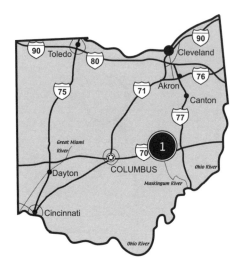

INFORMATION:
Ohio flint is known throughout the world for its brilliant colors. This durable variety of quartz cleaves easily, making it ideal for fashioning tools by early inhabitants of this area. Flint fashioned into tools has been found as far away as Kansas City. Early Europeans used this stone to sharpen tools and to grind grain. And of course, flint was used in flintlock rifles.

Chemical impurities account for the lovely varieties of green, pink, red, yellow, gray, white and black flint all intermingled.

Collecting is best done in the early spring before vegetation covers nice specimens. Be careful to avoid cuts on sharp flint edges. Wearing gloves and safety glasses is highly recommended.

MUSEUMS

Site 2.
Flint Ridge Museum
7091 Brownsville Road
Glenford, OH 43739
(740) 787-2476
http://www.ohiohistory.org/places/flint/
This museum, built at the site of a prehistoric quarry, contains fine examples of flint, including Native American artifacts and jewelry, as well as showing methods for mining and shaping flint.

Site 3.
Orton Geological Museum
Ohio State University
155 South Oval
Columbus, OH 43210
(614) 292-6896
This museum includes many dinosaur fossils including a Tyrannosaurus Rex.

Site 4.
The Cleveland Museum of Natural History
1 Wade Oval Drive, University Circle
Cleveland, OH 44106-1767
(216) 231-4600
http://www.cmnh.org/
Experience an earthquake, view a life-size reproduction of an 1890s mine, touch volcanic rock, and cook up rocks on a computer program by controlling the ingredients of time, temperature and pressure in the museum's Reinberger Hall of Earth & Planetary Exploration. Don't miss the exhibits of over 35,000 specimens including: rocks, minerals, ores, meteorites and gemstones.

Site 5.
Karl E. Limper Geology Museum
Department of Geology
Miami University
Oxford, OH 45056
(513) 529-3220
http://casnov1cas.muohio.edu/glg/Museum%20files/limperhome.htm
This museum exhibits many varieties of fossils, meteorites, metal ores, sedimentary, igneous and metamorphic rock and a collection of minerals with over 2,500 specimens.

CAVES

Site 6.
Ohio Caverns
2210 East State Route 245
West Liberty, OH 43357
(937) 465-4017
http://www.ohiocaverns.com
This guided ninety minute tour covers a mile and a half of underground caverns and chambers. Opened to the public in 1897, visitors travel past exquisite living cave formations as they learn a bit of history and geology. Ohio Caverns offers a special historical tour as an extension of the regular tour. Picnic area and gift shop are available.

Site 7.
Olentangy Indian Cavern
1779 Home Road
Delaware, OH 43015
(740) 548-7917
http://www.olentangyindiancaverns.com
The underground river that cut through the limestone of this cavern left a maze of winding passages and spacious rooms. Ancient Indian lore, history and the cave's natural beauty entwine on this guided tour. The first white explorer in this cavern was a westbound traveler who found his stray ox dead at the bottom of the natural entrance. J. M. Adams carved his name at the cave opening in 1821, where it remains today.

Site 8.
Perry's Cave
P.O. Box 335
Put-in-Bay, OH 43456
(419) 285-2405
http://www.perryscave.com
Fifty-five feet below the surface lies the limestone cavern used for centuries by Native Americans, and thought to have been "discovered" in 1813 by Commodore Oliver Perry, hero of the battle of Lake Erie. Loads of formations line the ceiling. The cave hides an underground lake whose level rises and falls with that of Lake Erie. For sometime this lake water was considered a cure-all. Tours take twenty minutes.

Site 9.
Seneca Caverns
P.O. Box 595
Flat Rock, OH 44828-0595
(419) 483-6711
http://www.senecacavernsohio.comm
A series of passages leads visitors to an underground river. The guided tour explains the cave's discovery and exploration, as the tour travels past unique formations. This is one of Ohio's largest caves.

Site 10.
Zane Shawnee Caverns
7092 State Route 540
Bellefontaine, OH 45311
(937) 592-9592
http://www.zaneshawneecaverns.org/
The bones and seashells deposited millions of years earlier, when this area was sea bottom, are still visible as the pressed limestone, which forms this cave. Calcium carbonate forms including soda straw, speleotherms, ribbon draperies and cave pearls all await the discovery of visitors to Zane Shawnee Caverns. The cave pearls are found nowhere else in Ohio. Tours travel three-quarters of a mile past pools and into unique chambers to see the geologic marvels of cave formations.

"We live in a heroic age." Andrew Carnegie

Pennsylvania is famous for steel and coal, but in 1740 colonists mined iron and copper. Anthracite coal, discovered at the headwaters of the Schuylkill River, fueled the furnaces used to smelt the ore. Surface mining began and a canal system brought the coal to Philadelphia.

In the late 1700s salt was discovered in the Conemaugh-Kiskiminetas Valley. Well water was briny, and salt crystals formed as water boiled from pots. Locals began mining the salt. The process involved digging wells, pumping away the water and then boiling away the liquid until crystallized salt remained. Local coal fueled the fires and the town of Saltsburg appeared on the state map. For a time, Pennsylvania was the nation's leading producer of salt. Salt was vital in food preparation, allowing meat to be cured and stored. By the 1830s the salty water ran out and crystal salt was found elsewhere.

All the locals knew about western Pennsylvania's Oil Creek. Before the settlers arrived, Seneca Indians used the black crude as a salve, medicine and insect repellent. They skimmed the viscous fluid from the water's surface as it bubbled from the ground. Settlers learned of the oil's existence and uses from the Indians. In the 1800s the annoying black oil intruded into several of the productive salt wells, closing several mines. When possible, the useless oil was drained away or burned off the precious salt.

Samuel Kier was the first to try to sell the oil. In the 1790s he bottled the unrefined crude and sold it as a cure-all. Not surprisingly sales were poor, so Kier used a still to refine his product for lamp oil. This useless oil now had a use. Others began collecting the crude. Eventually people got tired of damning the creek and skimming off the slowly percolating oil, and they decided to drill a well. Now oil gushed from the earth and by 1857 Titusville was America's first oil boomtown.

By the turn-of-the-twentieth century, coal and oil were big business. America's factories needed coal to fuel the Industrial Revolution, and Pennsylvania answered the demand. But this fuel did not come cheap.

By the 1840s Pennsylvania's annual mining report recorded over 50,000 deaths from mining accidents. Fires, cave-ins and mishaps with machinery killed many men. Immigrants from all over Europe came to America looking to improve their position. Many ended their lives in the mines. Unfair labor practices and dangerous conditions forced the men to organize. They demanded a shorter day, higher wages, and the right to buy goods at places other than the overpriced company stores. This early labor movement led to the formation of the United Mine Workers of America. It also led to one of the worst massacres on Pennsylvania's soil.

On Friday, September 10th 1897 a group of immigrant miners went on strike. The unarmed group marched peacefully from Harwood to Lattimer. The sheriff from Lattimer ordered them to halt, but most of the immigrants did not understand English. The sheriff's deputies fired on these defenseless men killing nineteen. The American newspapers condemned the slaughter. But this was not the last American tragedy to occur on Pennsylvania's soil.

In 1941 an explosion occurred in a mine in Harwick. One hundred and seventy-nine men were trapped in the shafts. Heedless of the danger, two of their fellow miners rushed back into the mine in a rescue attempt. These men died attempting to save their fellows, leaving behind wives and orphaned children. Pennsylvania industrialist Andrew Carnegie was so moved by the story, he funded a trust with five million dollars to recognize "civilization's heroes" and provide aid to those disabled or left behind by such individuals. The Carnegie Hero Fund Commission still exists awarding grants, scholarships and assistance. This group defines a hero as "A civilian who knowingly risks his or her own life to an extraordinary degree while saving or attempting to save the life of another person." On October 5, 2001 the Carnegie fund presented $100,000 to honor those heroes who died in the terrorist attacks of September 11, 2001, including those who died on Pennsylvania's soil in the plane crash while trying to regain control of the aircraft. Their actions likely saved many innocent lives. As Mr. Carnegie said, "We live in a heroic age."

Pennsylvania's mining history lives on in the workings of her people who live quiet heroic lives.

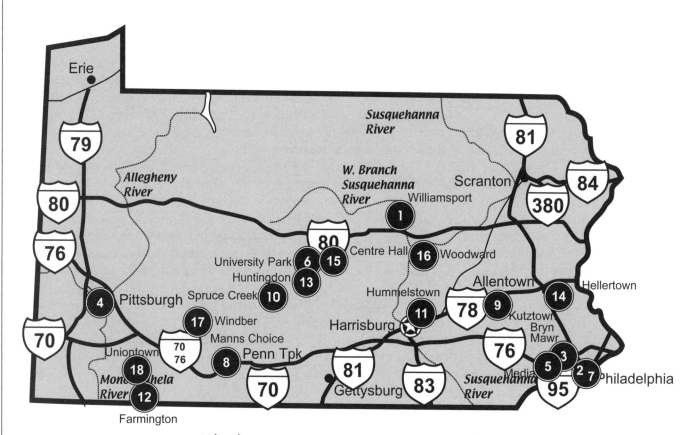

Diggings
1. Crystal Point Diamond Mine–Quartz Crystals

Museums
2. The Academy of Natural Sciences
3. Bryn Mawr College Vaux Mineral Collection
4. Carnegie Museum of Natural History
5. Delaware County Institute of Science.
6. Earth & Mineral Sciences Museum and Art Gallery
7. Wagner Free Institute of Science

Caves
8. Coral Caverns
9. Crystal Cave
10. Indian Caverns
11. Indian Echo Caverns
12. Laurel Caverns
13. Lincoln Caverns
14. Lost River Caverns
15. Penn's Cave
16. Woodward Cave

Points of Interest
17. Windber Coal Heritage Center
18. The Coal & Coke Heritage Center

ADDRESS:
c/o Raytowne
Box 1
1307 Park Avenue
Williamsport, PA 17701
(570) 323-6783
rpsmith@uplink.net
http://www.raytowneinc.com

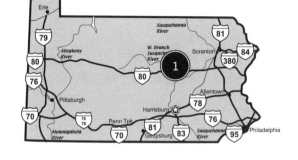

DIRECTIONS:
Take U.S. Highway 15 north from Interstate Highway 80 to Williamsport. Head into Williamsport on Fourth Street west to Rose Street. Turn right onto Rose Street and travel 2 blocks north to Park Avenue. Stop at the large brick buildings of Raytowne on the right just past Cemetery Street. Raytowne is a converted factory that is now a dance club and shopping center. From here you will follow the owner, Mr. Smith, down a rough jeep trail to the mine.

SEASON:
May 1 to October 31 (depending on the weather)

HOURS:
9:00 a.m. to 5:00 p.m. (by appointment only)

COST:
$40 – Adults
$20 – Children ages 12 and under

WHAT TO BRING:
Bring basic collecting tools and your own food and drink.

INFORMATION:
Quartz crystals found here vary in size from small, half-inch crystals to fifty-pound crystal clusters. A two hundred-yard long by twelve-feet deep trench has been dug to expose the quartz vein without damaging the crystal deposits. Crystals can be dug using picks, shovels and other hand tools. Bring a bucket and gloves. The crystals found are often clear, but sometimes smoky, orange or pale yellow. Larger crystals are single-terminated crystals. Smaller crystals may be double-terminated.

Mr. Smith tells us that you need to work carefully. Crystals are easily damaged by picks and shovels. A small garden claw or screwdriver are recommended for searching the red clay for crystals. Patience may be rewarded by the discovery of large, whole crystals.

Museums

Site 2.
The Academy of Natural Sciences
1900 Benjamin Franklin Parkway
Philadelphia, PA 19103
(215) 299-1000
http://www.acnatsci.org/
This museum includes extensive fossil exhibits and a permanent exhibit of crystals and gems.

Site 3.
Bryn Mawr College Vaux Mineral Collection
Park Science Building
Bryn Mawr College
101 North Merion Avenue
Bryn Mawr, PA 19010-2899
(610) 526-5000
610-526-5115
http://www.brynmawr.edu/Acads/Geo/mineralCollection/mincoll.html
This college has spent one hundred years collecting rocks, minerals, and fossils for use in the classroom and for exhibit and now has over 25,000 specimens in their collection.

Site 4.
Carnegie Museum of Natural History
4400 Forbes Avenue
Pittsburgh, PA 15213
(412) 622-3131
http://www.carnegiemuseums.org/cmnh/
This museum includes the Hillerman Hall of Minerals & Gems including 23,000 specimens and 2,300 gems.

Site 5.
Delaware County Institute of Science
11 Veterans Square
Media, PA 19063
(610) 566-5126
http://www.delcohistory.org/dcis/
This museum has a collection of Pennsylvania specimens in an exhibit entitled Mineralogy Of Pennsylvania. Additional specimens from the western United States are on display.

Site 6.
Earth & Mineral Sciences Museum and Art Gallery
112 Steidle Bldg., Pollock Road
University Park, PA 16802
(814) 865-6427
http://www.ems.psu.edu/Museum/
This university museum includes over 22,000 specimens of rock, mineral and fossil. Additional exhibits include glass, ceramics, metals, plastics, synthetic material, scientific equipment, mining equipment and archaeological relics.

Site 7.
Wagner Free Institute of Science
1700 West Montgomery Avenue
Philadelphia, PA 19121
(215) 763 6529
http://www.pacscl.org/wagner/about.html
Located in an 1865 Victorian building, this museum contains over 100,000 specimens, including a mineral and fossil collection. A series of lectures is also available to the public on a yearly schedule.

CAVES

Site 8.
Coral Caverns
Manns Choice, PA 15550
(814) 623-6882
http://www.coralcaverns.com/
Coral Caverns is the only known coral reef cavern in existence. This unique tour includes fossilized sea creatures and coral preserved in stone of what was once an inland sea.

Site 9.
Crystal Cave
963 Crystal Cave Road
Kutztown, PA 19530
(610) 683-6765
http://www.crystalcavepa.com
The most visited cave in Pennsylvania has many interesting formations, tours, gift shop, food, miniature golf, a rock shop, and nature trails.

Site 10.
Indian Caverns
HCR 1 Box 76
Spruce Creek, PA 16683
(814) 632-7578
http://www.indiancaverns.com
The cave was originally used by Native Americans and thereby gets its name. Native American artifacts and a picture table found in the cave are on display here. The cave offers stalactites, stalagmites and limestone formations illuminated along the mile-long concrete and gravel walkways. Picnic area, restrooms and a gift shop are available.

Site 11.
Indian Echo Caverns
368 Middletown Road
Hummelstown, PA 17036
(717) 566-8131
http://www.indianechocaverns.com
Flowstone, stalactites, stalagmites and underground lakes are viewed during the 45-minute guided tour. Afterwards guests can visit the gift shop or pan for gemstones at Gem Mill Junction.

Site 12.
Laurel Caverns
Route 40 West Caverns Road
Farmington, PA 15437
(724) 438-3003
http://www.laurelcaverns.com
This mountain top location offers a spectacular view of the surrounding area before you head underground to enjoy an illuminated guided tour. This cavern also offers spelunking trips for the more adventurous.

Site 13.
Lincoln Caverns
Rural Route 1
Box 280
U.S. Route 22
Huntingdon, PA 16652
(814) 643-0268
http://www.lincolncaverns.com
Unique flowstone formations abound, including Whispering Rocks. Take the one-hour tour through two separate caverns. Nature trails, workshops on geology and cave conservation are offered. Picnic areas, restrooms and a gift shop are available.

Site 14.
Lost River Caverns
P.O. Box M
Hellertown, PA 18055
(610) 838-8767
http://www.lostcave.com
Guided tours travel through five chambers, past crystal formations and an underground river. Visit the natural history rock and mineral museum, gift shop, lapidary, gems and indoor tropical garden. Restrooms and a picnic area are available.

Site 15.
Penn's Cave
RD 2, Box 165A
Centre Hall, PA 16828
(814) 364-1664
http://pennscave.com/
This location is the only all water cavern and a wildlife sanctuary. Open since 1885 and listed on the National Register of Historic Places, the cavern operators take visitors on a one-mile guided motorboat tour past flowstone and colorful limestone formations to Lake Nitanee. Penn's Cave also offers guided wildlife, farm and nature tours. Gift shop, antique shop, snack bar and picnic areas are available.

Site 16.
Woodward Cave
P.O. Box 175
Woodward, PA 16882-0175
(814) 349-9800
http://www.woodwardcave.com
This cave is still growing and features the Tower of Babel–the largest stalagmite in Pennsylvania along with many unique formations seen in five caverns. A gift shop, restrooms and picnic area are available.

Points of Interest

Site 17.
Windber Coal Heritage Center
501 15th Street
Windber, PA 15963-0115
(814) 467-6680
http://www.westsylvania.org/windber/frames.htmm
The heritage of Pennsylvania coal mining is exhibited at this facility. Staff dressed in nineteenth century mining costumes take groups through the center showing off the tools of the trade and demonstrates the uses of many of these items. The process of mining is explored as well as the life of miners. Washhouses and their importance in organizing labor and some of the first " lockers" developed for miners are exhibited.

Site 18.
The Coal & Coke Heritage Center
P.O. Box 519
Uniontown, PA 15401
(724) 430-4158
http://www.coalandcoke.org/
Collections of artifacts are now displayed at the Coal and Coke Heritage Center. Tools, taped interviews, documents, music and artwork all combine to give visitors an idea of the life of coal miners in this area. The center holds lectures, workshops and tours.

Pirates!

Rhode Island is a small state with a large spirit. As early as the 1630s this place became a colony promising religious freedom when English clergyman Roger Williams founded Providence. By 1639 Rhode Island had America's first Baptist church. Quakers followed in 1657, Jews arrived in what is now Newport in 1658 and French Huguenots immigrated in 1686.

Because of the proximity to the ocean, the sea and commerce became Rhode Island's first industry. The American Colonies had no navy; so private vessels were granted permission to attack England's enemies. This permission was called "Letters of Marquee." Vessels carrying these letters were free to attack and seize all cargo from enemies of the Crown. In the Colonies they were called privateers. To the vessels they attacked, they were pirates. Ten percent of the booty they stole, or confiscated depending on your point of view, was given to the Crown. The rest belonged to the vessel's crew, who took all the risk.

By the dawn of the 1700s, Rhode Island had only a little over 7,000 residents. In 1717 a famous privateer, Samuel Bellamy, sailed from Bahamas to Cape Cod. One of his captains, Palgrave Williams, detoured his small sloop to Rhode Island to visit his mother and sisters, and so escaped the storm that sank the Whydah and took her captain and most of her crew to their deaths. There is a museum on Cape Cod dedicated to the Whydah and its salvage.

By the 1760s Letters of Marquee were no longer granted by England, they were issued by Rhode Island. This state felt the tyranny of England's taxation policy most keenly as they lived and died by their imports and privateer's treasure. England's taxes strangled Rhode Island's economy. In 1772, far before the Boston Tea Party, Rhode Island colonists burned the British ship HMS Gaspee after intentionally luring it onto a shoal. They captured her crew and executed the captain. Every June the citizens of Warwick, Rhode Island celebrate this blow against tyranny during Gaspee Days.

Rhode Island again proved to be way ahead of their time, in 1774 by prohibiting further importation of slaves. In 1776, Rhode Island was the first colony to renounce ties with England and in 1784 they passed an emancipation act freeing slaves. One has to admire a plucky little state that spit in the King's eye and made freedom of religion and freedom for all men, regardless of color, such a top priority. They do not have mineral wealth. Rhode Island's wealth springs from her people.

With such a long and rich past, Rhode Island provides an interesting spot for metal detecting. Check local regulations for beaches and always ask permission when detecting on private land.

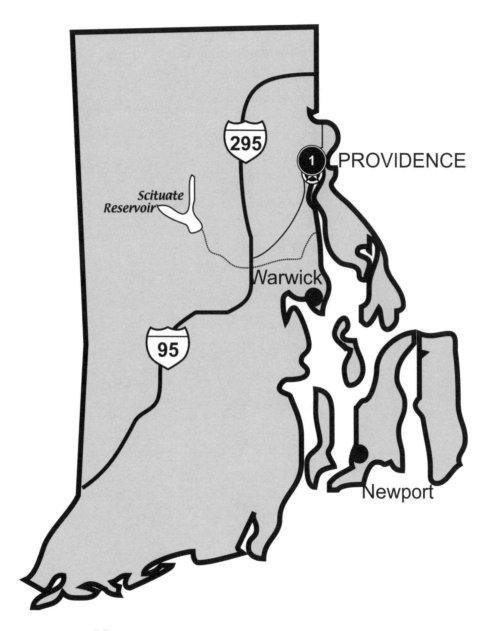

Museums

1. Museum of Natural History and Planetarium

MUSEUMS

Site 1.
Museum of Natural History and Planetarium
Roger Williams Park
Providence, RI 02905
(401) 785-9450
http://www.osfn.org/museum/
Among the many exhibits, visitors will find a collection of 15,000 rocks, minerals and fossils, including 350 million year old fossils from Rhode Island's Coal Age.

Clay pots in which to be buried.

One only has to look at the coastal plain of South Carolina to realize that the sea once dominated much of this landscape. The waters retreated leaving huge quantities of sand behind, sand, which provides raw materials for a number of industries. The beaches and rivers of South Carolina contain fossil remains. Farther inland the familiar rolling hills of red clay were once high mountains taller than the Himalayas. The Piedmont range gets its distinctive red color from the ferrous content of the clay. At one time the iron was painfully extracted for use in the Revolutionary and Civil Wars. Today it is the clay, itself that has real value and is used to make brick. South Carolina also has distinctive gray clay of very fine quality that is used by artists and potters. Ancient people of this area used the clay over 4,000 years ago making them among the first in the country to do so. Vessels were used for cooking and to store resources. Burial sites reveal clay pots were left with the bodies of the dead. Interestingly, pots were also used as coffins for children. This same clay was mixed with mud and used as plaster for the exterior surface of huts constructed of logs and woven sticks. Plainly this clay held great importance to these cultures.

Beachcombers in South Carolina are often rewarded with fossil finds. Edisto Beach State Park is a good place to search. Winter storms and the erosive action of the sea often bring fossilized shells, along with their modern counterparts, to the beach. More adventurous fossil hunters can turn to South Carolina's muddy rivers. Large fossilized sharks teeth are recovered by those who are willing to dive in water with nearly no visibility and inhabited by alligators. The rewards are great, fossilized teeth as large as six-inches can be found. Both Cooper River Divers and Deep South Rivers take divers fossil hunting.

If alligators and scuba diving is not your thing consider staying on dry land and mining in the northern region of South Carolina. At William's Property visitors can dig for amethysts. Other gemstones known in South Carolina include: aquamarine, emerald, yellow beryl, sapphire, garnet, topaz and tourmaline. This state also has gold.

Europeans first found gold in the 1800s but the metal was known by the Native Americans in the area long before. This discovery caused a rush and by the 1820s gold mining accounted for over 300,000 ounces recovered. Fools gold also prevailed. Glittering bits of mica pervaded the gold bearing earth. It was not until the much later that this annoying nuisance was recognized as a valuable resource. Mica is now mined and used in a huge range of products and processes. Mica, called vermiculite is added to potting soil to aid in drainage and aeration. It is used to stonewash denim to give the fabric a lived-in feel and mica is used in a number of filtration systems to name only a few of this versatile stone's applications. Those interested in searching for gold or mica should try the northeast portion of this state.

Finally treasure hunters should not overlook South Carolina's history. As one of the original thirteen colonies this state has old towns, churches, plantations and has endured the Revolutionary War, several Indian Wars and the American Civil War. Savvy metal detector hobbyist will find a plethora of potential search areas in South Carolina.

SOUTH CAROLINA SITE MAP

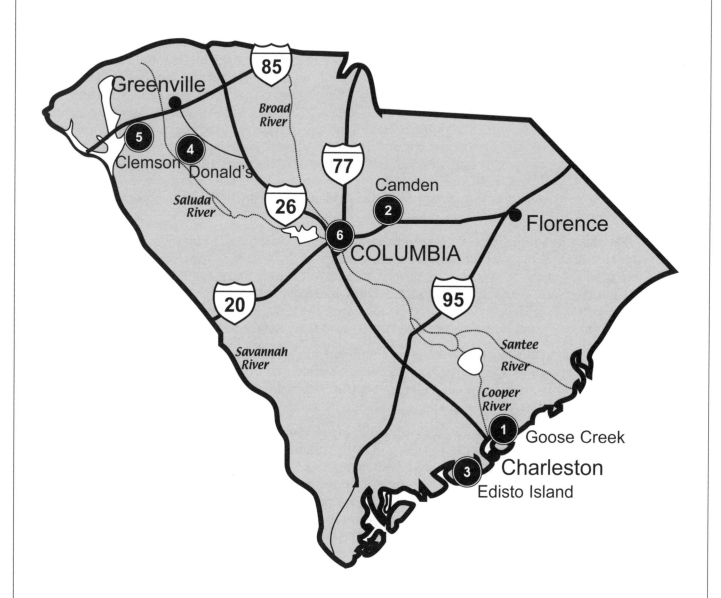

Diggings
1. Cooper River Divers
2. Deep South Rivers
3. Edisto Beach
4. William's Property

Museums
5. Clemson University Geology Museum
6. South Carolina State Museum

ADDRESS:
Cooper River Divers
260 Amy Drive
Goose Creek, SC 29445
(843) 572-0459
http://www.findaguide.com/frames/2410.htm

DIRECTIONS:
Call for directions to departure site.

SEASON:
Open all year round

HOURS:
Dive trips depend of the tides and generally run prior, after or during a slack tide period when flow is slowest. Call for specific times.

COST:
$55 to $65 per outing, which includes two dives.
Dive equipment rentals are available for an additional charge

WHAT TO BRING:
Weights, tanks, gear, lunch, dive equipment, collecting bag, hand tools such as a garden trowel and a South Carolina Hobby Collector's license.

INFORMATION:
Cooper River is well known for the fossilized Megalodon shark teeth, up to six-inches in length, that are recovered here. In addition, visitors with Cooper River Divers have found artifacts such as arrowheads, lances and spear points, pottery, antique bottles and fossilized bone from whale, mastodon and mammoth.

The river can have strong currents and visibility is generally about ten feet. In winter, visibility increases as the temperature drops. In the summer the water is warmer, but visibility decreases with the algae bloom.

ADDRESS:
Deep South Rivers
4679 Red Hill Road
Camden, SC 29020
(803) 428-3884
http://www.deepsouthrivers.com

DIRECTIONS:
Call for directions to departure site.

SEASON:
Open all year round

HOURS:
Dive trips depend of the tides. Call for specific times.

COST:
Two days of two-tank dives cost $390 per day
Two days of three-tank dives cost $450 per day
Fee includes: dive boat, captain and divemaster.

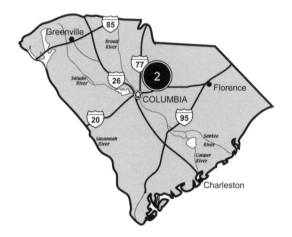

WHAT TO BRING:
Weights, tanks, gear, lunch, dive equipment, collecting bag, hand tools such as a garden trowel, and a South Carolina Hobby Collector's license.

INFORMATION:
   Deep South Divers promises a world-class fossil and artifact hunting dive. They put you in a productive area, known to yield fossils and artifacts and set you loose. Currents can be strong, visibility is typically about five feet and dive depth ranges from twelve to fifty feet. Dives are one to two hours in length.
   Finds recovered by divers include giant sharks' teeth, fossils from whale, bird, beaver, porpoise, camel, sloth, horse, alligator, mastodon, llamas, tapir and wooly mammoth. Artifacts recovered include bottles, pipes, buttons, spear points, coins and pots.

ADDRESS:
Edisto Beach State Park
8377 State Cabin Road
Edisto Island, SC 29438
(843) 869-2756
http://www.discoversouthcarolina.com/sp/spproduct.asp?PID=1298

DIRECTIONS:
From Interstate Highway 95 take Exit 57 to
SC-64 East to Jacksonboro. Turn left on U.S.
Highway 17 and drive 7 miles to SC-174. Turn
right onto SC-174 and drive to Edisto Beach.

SEASON:
Year Round

HOURS:
Open daily 9:00 a.m. to 4:00 p.m.

COST:
Nominal entrance fee is change at state parks.
For regular visitors a $40 pass entitles
holders free admission for the entire year.

WHAT TO BRING:
Bring beach shoes, a collecting bag, sunscreen, beach gear, a hat, food and drinks.

INFORMATION:
    Edisto beach has a saltwater marsh on one side and the Atlantic Ocean on the other. The park offers cabins, campsites, a playground, swimming, nature trails, picnic facilities, a nature center and a gift shop.

    Three million year old sediments mingle with new shells and sand on this lovely barrier island between Atlanta and Charleston. Beachcombers can begin at Edisto Beach State Park and walk north for about a mile to Jeremy Inlet. Fossil hunters walk the beach with an eye to the ground or stand in shallow water and screen material. Sharks' teeth, stingray tooth plates, bone and occasionally fossilized ivory are found here.

ADDRESS:
Lorraine Williams
Route 2, Box 259
Donald's (Due West), SC 29638
(864) 379-2148

DIRECTIONS:
The William's property is located on U.S. Highway 178. From Donald's, South Carolina, take State Highway 184, 4.1 miles to Due West, South Carolina. Turn left on State Highway 20 and go 1.2 miles, then take a left onto Ellis Road. Follow Ellis Road to the first brick house on the right.

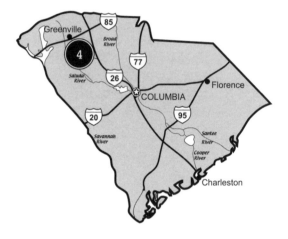

SEASON:
April to October

HOURS:
8:00 a.m. to dusk

COST:
$6 per day – Fee

WHAT TO BRING:
You will need a shovel, sifting screen with a quarter- to half-inch mesh, buckets and a closed container for your amethysts. No food is available so bring food and drinks.

INFORMATION:
This is a family-owned business, located in Abbeyville County, South Carolina. If you are planning a trip to the William's property to look for amethyst crystals, please call ahead two to three days in advance to make reservations. The digging is easy, so children can participate.

Some very nice amethyst crystals are found on the property. The amethyst is usually lilac or light purple in color. Occasionally a deep purple specimen is unearthed. It is possible to find small crystals lying on the surface in tailings piles, trails or open areas. Using a sifting screen can also prove productive. More ambitious miners look for quartz veins and use a crowbar and pick to access additional crystals.

Keep in mind that the most accessible spots have been picked over. To recover better specimens you will need to locate out of the way spots or overlooked areas.

MUSEUMS

Site 5.
Clemson University Geology Museum
Geology Museum
103 Garden Trail
Clemson, SC 29634-0130
864-656-4789
http://virtual.clemson.edu/groups/geomuseum/
This museum has minerals, fossilized plants and animals, samples you can touch and an exhibit of fluorescent minerals you can see glow under black lights.

Site 6.
South Carolina State Museum
301 Gervais Street
P.O. Box 100107
Columbia, SC 29202
(803) 737-4921
http://www.museum.state.sc.us/
A set of jaws and 43-foot model of a prehistoric great white shark hand over head at the South Carolina State Museum. The teeth from this extinct species are found in the rivers of South Carolina.

Million dollar mussels

Tennessee, the sixteenth state to join the union, had a rocky history and endured Indian Wars until well into the 1800s. This state seceded from the union and is the site of the bloodiest battle of the civil war, the Battle of Shiloh where 23,000 American's died. When Lincoln was assassinated it was Tennessee's son, Andrew Johnson, who became president. During the reconstruction period following the Civil War, Tennessee suffered with a large number, of residents unemployed and homeless. Industry was desperately needed to fuel this states failing economy. The industry came, but at a terrible price.

In the 1840s copper, which had been mined by the Cherokee for many years, was "discovered." Mining began in the southeastern corner of the state in the 1850s. A museum and mine in Ducktown, TN tells the story. Because there was no railroad system yet, communities grew around the mine. No coal could be imported by rail, so the forests were clear-cut to provide fuel for smelting the copper, iron and zinc. Over 30,000 acres were cleared and the smelting process pumped sulfur dioxide into the air. In 1878 mining stopped, only to resume when the railroad arrived to provide needed coal. Without the forest to hold the soil, rainstorms washed away the topsoil and the earth began to look as barren as the surface of the moon. In the 1930s the process of correcting this ecological nightmare began.

Gold is also found in Tennessee and was known about as early as 1799 in the area called Coker Creek. Modern prospectors can still pan for gold on Coker Creek. The rivers of Tennessee hold other unexpected riches, hidden away in the most unlikely of places.

Oysters were used for bait, hog food and occasionally people food. When Charles Bradford and James Johnson opened an oyster in the 1880s they only wanted something to put on their hook. But inside their oyster was a freshwater pearl. These pearls come in all colors and in various shapes including baroque, pear-shaped, spherical and irregular. Until this point pearls in Tennessee were unknown to settlers. The Native Americans likely were aware of the lustrous treasure within the mussels as they used shells for tools and ornament. The pearl was shipped off to Tiffany's in New York. Imagine their surprise when they received a check for $83 from the gem company. Several similar discoveries signaled the beginning of an industry and the near extinction of fresh water mussels. People from all over the state began collecting mussels and searching for pearls. When techniques were developed to cut buttons from the pearly inner surface of the mussel shell, the problem grew worse. The building of dams slowed the water feeding the mussel beds adding to the growing problem. Soon the shallows were barren of all mollusks. By the turn of the century, scientists were concerned. The invention of plastic buttons may well have prevented extinction of this species in Tennessee. The mollusk now live in deeper water where they are more difficult to harvest. Water pollutants still threaten the mussels.

In the 1960s the Japanese perfected a technique to introduce an artificial irritant to saltwater oysters and induce them into making a pearl. It happened that Tennessee mussel shells were perfect for the creation of the pellets inserted into the oysters. Harvesting began again and continues. Of the 127 species of mussel known a century ago – only seventy remain.

Most mussels are harvested by divers or with a technique called brailing. The work is dangerous and exhausting, but a good diver can earn a good wage. This industry brings over $20 million dollars annually to Tennessee. New regulations are in place to prevent harvesting mussels below a certain size to prevent the depletion of this state's valuable resources. Pearl enthusiasts will enjoy A Pearl of A Tour in Camden Lake, which includes a pontoon boat ride, barbecue and historical tour. There is also a pearl museum and a jewelry showroom here.

Diggings
1. Burra Burra Mine–Garnet and Pyrite
Museums
2. Frank H. McClung Museum
Caves
3. Appalachian Caverns Foundation
4. Bristol Caverns
5. Cumberland Caverns
6. Forbidden Caverns
7. Lost Sea
8. Raccoon Mountain Caverns
9. Ruby Falls
10. Tuckaleechee Caverns
Points of Interest
11. Pearl Tour

ADDRESS:
Burra Burra Mine Historic Mine Site
Ducktown Basin Museum
P.O. Box 458
Ducktown, TN 37326
(423) 496-5778
http://www.gamineral.org/commercial-burra-burra.htm

DIRECTIONS:
Ducktown is two hours and 101 miles from Atlanta. From the Interstate Highway 75 and Interstate Highway 575 north interchange, take Interstate Highway 575 north and follow the four lanes as the turn into SR 5 at Nelson and then U.S. Highway 76 (SR 5, SR 515) at East Ellijay. Turn left onto SR 5 at the McDonalds in Blue Ridge. Continue to Mccaysville. SR 5 becomes SR 68 at Copperhill, Tennessee. Follow SR 68 to Ducktown and watch for the sign.

SEASON:
Open year round

HOURS:
November to April- 9:30 a.m. to 4:00 p.m. Monday through Saturday.
May to October - 10:00 a.m. to 4:30 p.m. Monday through Saturday.

COST:
Museum admission and collecting fees are $3.00 for adults, $1.00 for teens and 50 cents for children under twelve.

WHAT TO BRING:
Bring a small sledgehammer, chisel, rock hammer, garden trowel, five-gallon buckets, safety glasses and a bag or container to carry finds. A hand truck or cart is recommended to help carry heavy pyrite once your buckets are full.
You will be scrambling on rock piles, so proper footwear is a must. Bring water, all your food, sunscreen, hat and gloves.

INFORMATION:
　　Ducktown, known for duck hunting, is also the site of the Burra Burra Mine and Heritage Museum. Once a booming copper town in the mid-1800s, this state owned historic site is now enjoyed by many rock collectors. Over 300 acres of denuded land is set aside to remind visitors of the devastation of the mining process in this area. At the museum you will find displays on copper mining and local history. After a visit to the Museum, drive a short distance to see the massive machinery, standing like rusting dinosaurs beside the central headframe. Collecting is permitted on the ground around the central headframe.

Museums

Site 2.
Frank H. McClung Museum
The University of Tennessee
1327 Circle Park Drive
Knoxville, TN 37996-3200
(865) 974-2144
http://mcclungmuseum.utk.edu/
Exhibits here include anthropology, archaeology, local history, natural history and decorative arts – showcasing Tennessee's past in geology, art and culture from prehistory to modern times. The exhibit entitled Geology and Fossils of Tennessee includes rocks, fossils (including the only dinosaur ever found in this state, a duck-billed dinosaur called a hadrosaur), minerals and geologic features to illustrate the state's history.

Caves

Site 3.
Appalachian Caverns Foundation
420 Cave Hill Road
Blountville, TN 37617-6224
(423) 323-2337
http://www.appalachiancaverns.org/
Native Americans, earlier settlers and moonshiners knew about this cavern long before it was ever opened to the public. Electric lights, dry walkways and knowledgeable guides make a visit today much more enjoyable. Visitors will see many corridors and rooms. One chamber soars to 135 feet. Manganese, iron, copper and calcium provide the color for the many exciting formations within this cave system.

Site 4.
Bristol Caverns
P.O. Box 851
Bristol, TN 37621
(423) 878-2011
http://www.ohwy.com/tn/b/briscave.htm
Native Americans used the cave as an escape route. Visitors walk along the banks of an underground river and past exciting geologic formations.

Site 5.
Cumberland Caverns
1437 Cumberland Caverns Road
McMinnville, TN 37110
(931) 668-4396
http://www.cumberlandcaverns.com
Boasting to be one of the largest show caves in the state, Cumberland Caverns has some of the largest underground chambers in the country. Waterfalls, pools and amazing formations await visitors. Tours run May to October.

Site 6.
Forbidden Caverns
455 Blowing Cave Road
Sevierville, TN 37876
(865) 453-5972
http://www.smokymtnforbiddencaverns.com/
Flint, arrowheads, scrapers and knifes are a few of the items left behind by the Native Americans who used this cave for shelter. A running stream flows through the cavern past some spectacular living formations. The largest known wall of onyx known in existence is also here.

Site 7.
Lost Sea
140 Lost Sea Road
Sweetwater, TN 37874
(423) 337-6616
http://www.thelostsea.com
This cave gleans its name from the largest underground body of water listed by Guinness Book of World Records, the world's largest underground lake. Divers have explored some of the lake, which is 220 feet wide and 800 feet long. But there is more to this cavern system. Cherokee used the cave and left artifacts behind including arrowheads, weapons, pottery and jewelry. A giant Jaguar from the Pleistocene period left a track deep within the cave before dying in the cave some 20,000 ago. The bones are now displayed in the American Museum of Natural History in New York City. An early settler used the cave as a root cellar and the saltpeter was mined during the Civil War for gunpowder. Glassbottomed boats carry visitors around the lake, which is stocked with large rainbow trout. Sorry, no fishing!

Site 8.
Raccoon Mountain Caverns
319 West Hills Drive
Chattanooga, TN 37419
(800) 823-2267
http://www.raccoonmountain.com/html/caverns.html
Fossils from the ancient sea, from which the limestone was formed, are still visible including coral, crinoids and other marine animals. Weak carbonic acid from ground water seeping into the caves cut the limestone leaving the caverns visible today. Tours take visitors through chambers and corridors, past unusual formations.

Site 9.
Ruby Falls
Ruby Falls
1720 South Scenic Highway
Chattanooga, TN 37409
(423) 821-2544
http://rubyfalls.com/
Visitors will enjoy a visit to Ruby Falls, named after a spectacular 145-foot waterfall. Guided tours are available.

Site 10.
Tuckaleechee Caverns
825 Cavern Road Box 381
Townsend, TN 37882
(423) 448-2274
http://www.ohwy.com/tn/t/tuckcave.htm
Tours of approximately one hour take visitors past high waterfalls, and onyx formations, for which the cave is noted, and by amazing limestone formations.

POINTS OF INTEREST

Site 11.
Pearl Tour
Birdsong Resort & Campground
255 Marina Road
Camden Kentucky Lake, TN 38320-9699
(731) 584-7880
http://www.birdsongresort.net/
This resort has much to offer, including a lakeside RV campground and recreational complex on the largest man-made lake in America. Birdsong also operates a pearl farm, pearl museum and jewelry showroom. Guided tours take between three and seven hours, explain the cultivation process and mussel industry and include a pontoon boat ride to the pearl farm.

Marble of the highest quality

It is hard to imagine Vermont as a wilderness, but that is what it was in the 1700s when the first European settlers ventured into its virgin forests. In 1768 a patch of snow was discovered not to be snow at all but a deposit of white marble. William F. Barnes started a kiln there to burn the marble into lime and later decided to quarry material for gravestones. He purchased the land for the price of one horse. The stone beneath this land was worth millions. By 1784 marble was quarried in Dorset to make sidewalks, fireplace mantles and tombstones. The upper layers of poor quality stone were gradually removed to reveal a hard, sturdy marble of the highest quality. This luminous stone graces the New York City Public Library, the U.S. Supreme Court and the Jefferson Memorial. Early quarrying was done with hand tools. Cut blocks of marble were hauled by oxen to market twenty-five miles away in Whitehall. The cost of this process cut deeply into profits. No rail system yet existed and Vermont did not have the river system that their neighbors in Connecticut used to haul their brownstone to market. Therefore they sold marble locally where they could. Marble was used in towns for sidewalks. Marble became as common bricks for building. Even modest homes of this period have marble floors.

By 1851 a rail line reached the quarry. The introduction of a new cutting method speeded production and allowed marble to be cut into uniform slabs. The "gang of saws" consisted of a soft iron strip that was rubbed across the stone by an engine. Sand was added and then water dripped into the crevice and the stone was cut. This method is ancient, but the addition of an engine greatly speeded the process and quality of the stone produced. Marble was now cut into blocks for statues and slabs for buildings. This white marble graces the state capitals of Idaho, Kentucky, Oklahoma, Oregon and South Dakota. Many of the union soldiers who died in the Civil war have tombstones from Vermont. Churches, libraries, universities, capitals and mausoleums all use this marble. Vermont marble is not limited to black and white. The stone forms in different colored layers including striped, gray, green, blue and pink.

By 1885 the quarries and mills in West Rutland employed 3000 workers and ran day and night. Over the years some quarries tapped out while others filled with water making today's swimming holes. Marble is still in demand.

In 1996 a stone called Verde Antique began to be removed from Windham County. It is mixture of serpentine, translucent light to dark green rock and white marble. This ornate stone is in high demand.

Prior to the American Revolution, in an area which is now part of Vermont, black marble was discovered and quarried. The Isle of La Motte's most distinctive black marble contains fossils throughout the stone. The site of a French fort in 1666, settled in 1788, this quarry was active through the late 1700s and supplied the floor for the new Vermont state capital and was used in the construction of Radio City Music Hall in New York City. The La Motte quarries are now local swimming holes on this island of Lake Champlain. The Fisk Quarry is open to the public for picnics, swimming and rock hunting.

In the 1850s a surprising discovery was made in Plymouth. Gold was found in the area called Buffalo Creek. Mills and crushers were hastily constructed and prospectors flooded in. By the 1880s the placer gold was exhausted and no veins were found. This gold likely came to Vermont carried in the glaciers during the Ice Age and is not indigenous to the region. Small amounts of quartz and gold have been found, but in such quantities as to make prospecting unprofitable.

About the same time copper was discovered. Unlike the gold, it was not glacier deposits. A copper boom occurred. In 1881, the Vermont Copper Company employed a little over 800 workers laboring to work the mines, smelting process and huge furnaces. Chimney flues ran up the hills, carrying away the noxious smoke. The smelter emissions resembled the odor of rotten eggs. The hills near the mine were described as being desolate of all vegetation. A century later nothing grows near the site of the smelters. Through mis-management and dwindling ore the mine floundered. In 1883, on June 29th, payday, the mine closed. Workers, who had gone months without pay, raided the company store and marked on the home of the owner in West Fairlee. The National Guard was summoned and 184 soldiers arrived in town on July 6th. Instead of a riotous crowd, they found a group of miner's representatives who met them and explained their grievances. The soldiers were so disturbed, they gave the miners their food and returned to their train. The event was called the Ely War. In the end the company went bankrupt and the miners received nearly nothing for their labors.

Modern gold miners are still seen in Plymouth. Hobbyists have claims along the streams. For those traveling through, gold sluicing and panning is permitted in the creeks and streams at Plymouth State Park. To visit a marble quarry, now swimming hole, travel to the Fisk Quarry at Isle La Motte in Lake Champlain. Those interested in the quarrying process and history of marble in this state should try the town of Proctor, which has the Vermont Marble Exhibit.

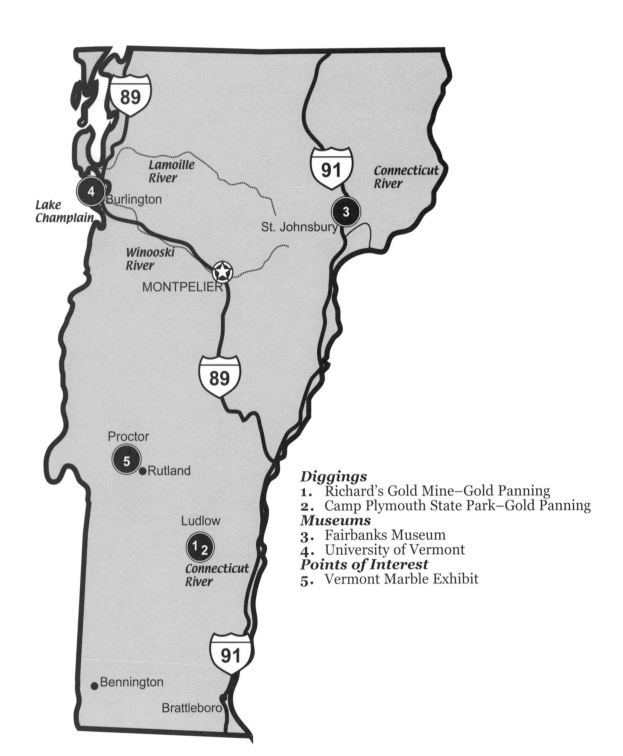

Diggings
1. Richard's Gold Mine–Gold Panning
2. Camp Plymouth State Park–Gold Panning

Museums
3. Fairbanks Museum
4. University of Vermont

Points of Interest
5. Vermont Marble Exhibit

ADDRESS:
1944 Main Street
Athol, MA 01331

DIRECTIONS:
The Richard's Gold Mine is located in
Ludlow, Vermont. Reservations are required
and directions are given at that time.

SEASON:
May to October – Weather permitting

HOURS:
11:00 a.m. to 5:00 p.m.

COST:
$25 per person per day – Fee
$15 per person per day – Fee for groups of
nine or more

WHAT TO BRING:
You will be supplied with everything you need to pan for gold including: gold pans, vials to hold
your finds, shovels and picks. There are no food or drinks available at the mine, so bring them
with you.

INFORMATION:
Richard's Gold Mine was first worked in the late 1890s, when gold was discovered by a
prospector returning from the California gold fields. Mining operations were conducted
through the 1940s. Large amounts of placer gold and nuggets were collected.

During World War II, the War Powers Act closed all mines not essential to the war effort.
This act also made it illegal for foreigners to own mines in the United States. This forced the
Canadian owner to abandon his mine.

The mine is located in the central Green Mountain area. Gold found here ranges from small
nuggets to gold dust. Only gold panning is permitted. You may not use dredges or other mining
tools. Richard's offers gold panning classes, trips and lectures.

ADDRESS:
Camp Plymouth State Park
Route 1, Box 489
Ludlow, VT 05149
(800) 299-3071 or (802) 228-2025
parks@fpr.anr.state.vt.us
http://www.vtstateparks.com/htm/plymouth.html

DIRECTIONS:
From Ludlow, Vermont take State Highway 100 north. Turn right at the Echo Lake Inn where the sign points to Camp Plymouth State Park. Drive 1.5 miles around the lake and take the first left at the sign to Camp Plymouth State Park. The park is half a mile after the left turn.

SEASON:
Memorial Day weekend to August 30

HOURS:
10:00 a.m. to dusk

COST:
$2 – Adults
$1.50 – Children

WHAT TO BRING:
Gold mining equipment and bug spray. May is blackfly season in Vermont.

INFORMATION:
The first gold found in Vermont was found in Buffalo Brook in what is now Camp Plymouth State Park. A gold nugget was found by an unknown fisherman in 1826 and the search for gold was on in Vermont. By 1885, the cost of mining was more than the gold yielded here and commercial mining ceased. If you walk half a mile up the Echo Lake Overlook Trail you can see the remains of an old collapsed air shaft, off to your left after the walkover bridge. This is all that remains of the gold mine. Buffalo Brook is on your right and this is where panning is permitted.

The park allows gold dredging and gold mining without a permit, but you will need to call ahead to notify the ranger station prior to your arrival. You may also metal detect for nuggets, but no artifacts may be removed, according to Vermont state law. Dredging requires a state permit be obtained by writing or calling the Stream Alteration Engineer, State of Vermont, Water Quality Division, 450 Asa Bloomer State Office Building, Rutland, Vermont 05701, (802) 786-5906.

The walls of the brook are steep. Bedrock is exposed in some places along the brook and the brook is very shallow in other places. Much of the gold found here is fine flakes, but a few nuggets have been reported. Vermont has some native gold, however. Most of the gold here was deposited by glaciers during the last ice age.

MUSEUMS

Site 3.
Fairbanks Museum
1302 Main Street
St. Johnsbury, VT 05819-2248
(802) 748-2372
http://www.fairbanksmuseum.com/
This museum includes a natural science collection of mounted animals, shells, rocks and minerals.

Site 4.
University of Vermont
Department of Geology
Perkins Hall
Burlington, VT 05405-0122
(802) 656-3396
http://geology.uvm.edu/
This one room exhibit includes minerals, rocks of Vermont and the nation and various fossils. The highlight of the display is "Charlotte" the whale, whose skeleton was discovered 150 years ago in a field in Vermont. How did a whale end up in rural Vermont? Researchers believe Lake Champlain was once a much larger saltwater sea.

POINTS OF INTEREST

Site 5.
Vermont Marble Exhibit
52 Main Street
Proctor, VT 05765
(800) 427- 1396
http://www.vmga.org/rutland/vtmarble.html
Visitors will explore the history of the marble industry in Vermont, view a color film, watch sculptors, view historic photos and see marble displays. The geology room includes fluorescent minerals, dinosaur and other fossils. Don't forget to visit the gift shop for your own marble souvenir.

When the fairies heard the news, they wept.

Thomas Jefferson authenticated gold in Virginia in 1782. The ore sample was about four pounds of quartz laced with gold. The Native people collected gold long before earlier French and Spanish found the metal. These explorers and trappers noted in their reports that local Indians had gold in their possession in the 1500s. Commercial Mining began in the 1800s in Orange and Goochland counties using rocker boxes, pans, and some hydraulics. The best-known of these mines is the Tellurium. Early methods included crushing the ore by hand with iron mortars before extracting the gold. Later horses were used to help crush the stone in stamp mills. Much of these operations were abandoned when gold was discovered in California. The outbreak of the Civil War stopped all mining in Virginia.

This state is rich in collectable rock and gems. Garnets are found on the northeast side of the Tyle River near Arrington. Amelia County had two garnet mines at one time producing gem-quality stones. In 1991 a 2,800-carat stone was unearthed and named the Rutherford Lady. Virginia garnets are found in colors other than red. Varieties of orange, yellow and brown are common.

Amelia County also has amazonite in deep blue-green. Translucent amazonite is highly prized and also found here. The Morefield Mine allows visitors to dig for amazonite and other gems including quartz, amethyst, garnet and moonstone.

Moonstone is a gemstone of feldspar. Feldspar is found in abundance, being one of the most common minerals of the earth's crust. Gems are rare, however. Virginia uses its industrial feldspar in the manufacture of glass, plumbing fixtures, pottery and tile.

Another unique find in Virginia are the staurolite crystals otherwise called fairy stone, found in Stuart. Local legend says that the fairies that lived in this valley wept at the news of the crucifixion of Jesus Christ. Their tears fell to earth and crystallized to form small stone crosses. The fairies are gone but the stone tears remain. Whatever their origin, staurolite, from the Greek word for cross are a brown combination of silica, iron and aluminum. These substances materialized under great heat and pressure to create twin forms resembling crosses. Nowhere else in the world are these crosses found in such abundance or so perfectly formed. The crosses occur in several varieties.

The fairy tale of origin is charming. The history of belief surrounding these crosses is equally fascinating. They were and still are thought to provide the wearer with a plethora of benefits. Historically held in superstitious awe, they were believed to protect the wearer from witchcraft, illness, accidents and other disasters.

Staurolite, known as baptismal-stone, was worn as an amulet at baptism. It is said that President Teddy Roosevelt wore a fairy stone on his watch chain. Today this stone is still sought after for its believed benefits and purported powerful protection.

Fairy stones are easy to mine. Visitors to Fairy Stone State Park in Stuart can get tips from the rangers and see a nice collection of specimens. Digging is done in soft clay with a shovel or hand tools and you keep what you find.

Virginia is rich in history as well and is the sight of many battles fought on American soil, including the Revolutionary and Civil Wars. During these uncertain times civilians hid caches of valuables including coins and jewelry to protect them. Soldiers left artifacts along with their blood. This makes Virginia a treasure hunters dream and many unusual and valuable historic finds have been unearthed in Virginia. Metal Detector hobbyists must consider this state as a possible destination. Be mindful that detecting is not permitted on historic battlefields and some state parks. Know local regulations before you dig.

With a range of gems, gold and specimen rocks, Virginia has something to please all rockhounds and prospectors.

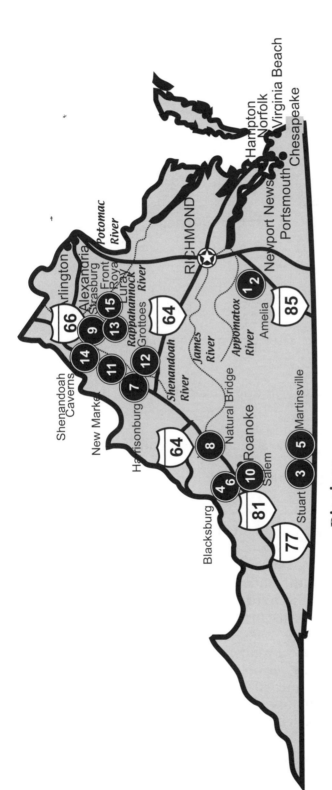

Diggings
1. Morfield Gem Mine
2. Deck Boyles Farm and Mine
3. Fairy Stones State Park

Museums
4. Virginia Museum of Natural History at Virginia Tech

Points of Interest
5. Virginia Museum of Natural History at Martinsville
6. Virginia Tech Geological Sciences Museum
7. James Madison University Mineral Museum

Caves
8. Caverns of Natural Bridge
9. Crystal Caverns at Hupp's Hill
10. Dixie Caverns
11. Endless Caverns
12. Grand Caverns
13. Luray Caverns
14. Shenandoah Caverns
15. Skyline Caverns

ADDRESS:
13400 Butlers Road
Amelia, VA 23002
(804) 561-3399

DIRECTIONS:
From Interstate Highway 95, take Exit 15A in Richmond, Virginia to Interstate Highway 195 to the Powhite Parkway, Virginian Route 76 (a toll road). Exit onto Virginian Route 288 and follow this toward Interstate Highway 95 South. Exit onto U.S. Highway 360 West (Hull Street Road) toward Danville. Follow U.S. Highway 360 west about 14 miles across the Appomattox River and into Amelia County. Drive 3.3 miles past the county line and turn left onto County Road 628. The mine entrance is 1 mile south on the left.

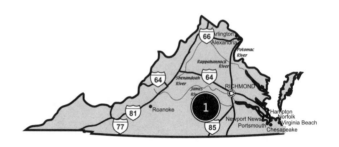

SEASON:
March through December (closed to the public after December 15 to March 21)

HOURS:
9:00 a.m. – 4:30 p.m. – Closed Sundays; Closed Mondays except in June, July and August

COST:
$10 – Adults (includes use of tools–buckets, rakes, screens, sluice and shovels)
$7 – Children ages 4 and up
Collectors are asked to limit themselves to a five-gallon bucket of material per day

WHAT TO BRING:
You need standard mining equipment, a lunch and plenty to drink.

INFORMATION:
    This is an ideal place for families. The flume and dig site are in close proximity and the terrain is gently sloped. The mine dumps contain hard-packed dirt. Visitors are allowed to dig anywhere within the marked area. The sluice has a canopy to keep you in the shade as you wash your dirt through the shifting screen. The mine has a gift shop, lapidary shop and indoor restrooms. Drink machines and a picnic area are a welcome sight when you need a break.
    Pay close attention to the sample minerals on display and listen closely to the owner's instructions. Refer back to the displays as you unearth your treasure trove.
    Amazonite is plentiful here and of good quality. But that's not all! There is garnet, quartz (amethyst, clear rutile and smoky topaz), calcite, chrysoberyl, fluorite, moonstone, mica, tourmaline and zircon–just to name a few!

ADDRESS:
13001 Butlers Road
Amelia, VA 23002
(804) 561-2395

DIRECTIONS:
From Amelia, Virginia, take Route 628/
Bulters Road approximately 3 miles. Look for
Mailbox #13001 on the right as the marker for
the Boyles Farm and Mine.

SEASON:
Open all year round

HOURS:
Daylight

COST:
$3 per day – Fee

WHAT TO BRING:
You need standard mining equipment, a lunch and plenty to drink.

INFORMATION:
The mine was first worked during the Civil War when prospectors looking for mica also came across green crystals of beryl. Mica was mined for its resistance to heat and transparency and was sometimes used in stoves and lanterns instead of glass. Mica is plentiful here.

The elusive and valuable beryl is more difficult to unearth. You may dig all day and only find a piece or two. Beryl unearthed can be as fine as pencil lead. Large specimens up to an inch in diameter and several inches long have also been found. Blue-green hexagonal crystals are found with the black mica in the red clay soil of the fifty-foot pit mine.

Beryl comes in a variety of colors. The beryl family includes tourmaline, emerald and aquamarine. The color of the beryl found here includes clear, gray, gray-blue, blue-green, brown, yellow and pale blue (aquamarine).

The mine uses a back hoe to uncover new material periodically. The mine area can be searched with shovel and screens. Use your screen to shift the red clay and uncover crystals. Beryl crystals grow in association with feldspar and mica.

Wear old clothing on your expedition. The red clay is especially hard on clothing and stains are often permanent.

ADDRESS:
967 Fairy Stone Lake Drive
Stuart, VA 24171
(276) 930-2424
revs@dcr.state.va.us
http://www.dcr.state.va.us/parks/fairyst.htm

DIRECTIONS:
Fairy Stone State Park is located in the foothills of the Blue Ridge Mountains in Patrick and Henry Counties. Access to the park is via State Highway 57 from Bassett or U.S. Highway 58, or State Highways 8 and 57 from the Blue Ridge Parkway.

SEASON:
Best collecting is in the Spring and Summer
The park is open year round

HOURS:
Daylight

COST:
$2 per day – Entrance fee
$18 per night – Camping fee
Free – Collecting at Fairy Stones

WHAT TO BRING:
You will need a shovel, hand rake, trowel and a pail or bucket to hold your fairy stones. A quarter-inch sifting screen may be helpful.

INFORMATION:
Fairy stones (staurolite crystals) can be found on the state park's property. These crystals form in various crosses and are very popular as religious necklaces. The access to the collecting area is left of the main park entrance on State Highway 57 at the first gas station on the left–"Haynes 57." The land at the left of the station is park land. A small parking lot is available. The State Parks Department requests that you only take a few samples for personal use.

The legend of the fairy stones is rather charming. It is said that local fairies lived and played in this glen. The fairies, upon hearing of the death of Christ, wept. Their tears crystallized into beautiful crosses. For years people have used these little crosses to protect against sickness, accidents and disaster.

This is the heart of Appalachia. The area has a strong tradition of crafts and music. The park is located in a very rural area.

Museums

Site 4.
Virginia Museum of Natural History at Virginia Tech
428 North Main Street
Blacksburg, VA 24061
(540) 231-3001
http://www.vtmnh.vt.edu/information.html
An exhibit titled "Minerals and Their Uses" includes gemstones, fossils, rocks and minerals, plus a section on rocks of Virginia.

Site 5.
Virginia Museum of Natural History at Martinsville
1001 Douglas Avenue
Martinsville, VA 24112
(276) 666-8600
http://www.vmnh.org/
This museum has specimens from Virginia including 13 miles of hard rock core.

Site 6.
Virginia Tech Geological Sciences Museum
Department of Geological Sciences
Virginia Polytechnic Institute and State University
4044 Derring Hall
Blacksburg, VA 24061-0420
(540) 231-6521
http://www.geol.vt.edu/outreach/museum.html
This museum is located in Derring Hall and includes a mineral collection, fossils and items of local mining history.

Site 7.
James Madison University Mineral Museum
Department of Geology
MSC 7703
James Madison University
Harrisonburg, VA 22807
(540) 568-6421
http://csm.jmu.edu/minerals/
Located on the top floor of Miller Hall, this collection contains 500 crystals and gemstones and fluorescent minerals.

CAVES

Site 8.
Caverns of Natural Bridge
P.O. Box 57
Natural Bridge, VA 24578
(800) 291-2121 or (540) 291-1896
http://www.naturalbridgeva.com
George Washington carved his initials on the wall of the bridge where he surveyed the Natural Bridges site in 1750. Thomas Jefferson owned the land in 1774 where he built a two-room retreat. By 1883 the area was made into a resort and it remains so today. Through the 1700 and 1800s the two most visited attractions in America were Natural Bridges and Niagara Falls. Today the 215-foot natural bridge and cavern draw countless visitors. The cavern is the deepest commercial cave in the east and includes many striking limestone formations.

Site 9.
Crystal Caverns at Hupp's Hill
33231 Old Valley Pike
Strasburg, VA 22657
(540) 465-8660
http://www.waysideofva.com/crystalcaverns/history.asp
Native tribes used this cave as shelter on hunting expeditions, judging from the artifacts found at this site. Discovered in 1750 and explored by the landowner who periodically invited his friends and neighbors to witness the fascinating formations, which he lit by lantern and candle. Those same marvelous formations are still there, but the illumination is much improved.

Site 10.
Dixie Caverns
5753 North Main
Salem, VA 24153
(540) 380-2085
http://dixiecaverns.com/
Discovered by a dog and several farm boys, this cave opened to the public in 1923 and offered views of this beautiful cavern's limestone formations including: draperies, columns, stalagmites and stalactites.

Site 11.
Endless Caverns
P.O. Box 859
New Market, VA 22844
(540) 896-2283
http://www.endlesscaverns.com
This cavern must have seemed endless to the two boys who discovered the entrance and explored with rope and candle. Explorers have mapped five miles of the cave to date. Guided tours take a little over an hour and includes geology and history. The cave formations are illuminated by white light only.

Site 12.
Grand Caverns
P.O. Box 478
Grottoes, VA 24441
(540) 249-5705
http://www.i81virginia.com/caverns.html
This cavern certainly matches its name and has pleased visitors since 1806. Cave formations will impress and so will the giant 280-foot-long Cathedral Hall, a chamber that rises 70 feet and is one of the largest in the east. Draperies, columns, stalagmites, stalactites and unusual shield formations are all awaiting visitors to Grand Cavern.

Site 13.
Luray Caverns
P.O. Box 748
Luray, VA 22835
(540) 743-6551
http://www.luraycaverns.com
Luray is the largest cavern in the east and is an underground world full of beautiful limestone formations including: draperies, columns, stalagmites and stalactites.

Site 14.
Shenandoah Caverns
261 Caverns Road
Shenandoah Caverns, VA 22847
(540) 477-3115
http://www.shenandoahcaverns.com/Caverns/
Natural occurring minerals help add color to the bacon formations, stalactites, stalagmites and flowstone which fills this cavern formed by water and a long ago earthquake.

Site 15.
Skyline Caverns
P.O. Box 193
Front Royal, VA 22630
(800) 296-4545
http://skylinecaverns.com//
This cave is justifiably proud of their anthodite, which are delicate formations with white spikes radiating outward in all directions. Growth of these anthodites is estimated at one inch per seven thousand years, but how they form is not known.

The companies cheated the miners with rigged cars and other ploys.

Three hundred million years ago what is now West Virginia was a Prehistoric swamp. Dead plant material fell to the water by the ton, gradually turning into peat and then coal. It is estimated that twelve inches of coal took over 10,000 years of accumulated vegetative material to create. West Virginia has coal deposits twenty-five feet thick in places and averages three feet.

The recession of the ocean left something else behind as well – salt. Beneath the ground, beds of salt are 50 feet thick in places. The mining process involved drilling to the salt level and pumping fresh water down into the shaft, then extracting the water carrying salt in solution. The water was later boiled away to recover the "white gold."

West Virginia has another valuable resource in the form of high quality sandstone. This stone makes excellent raw material for the production of glass. The presence of sandstone and natural gas made it possible for West Virginia to lead the nation in the production of glass in the 1800s, producing windows, bottles, pitchers, glasses and tumblers. Sandstone was also quarried in blocks for the construction of bridges, paving and buildings. Another type of sandstone, called abrasive stone, was popular for grinding wheels and sharpening stones. The use of these stones has fallen off with the development of new abrasives to serve the same purpose.

But West Virginia's most active mining industry is coal. Discovered in 1742, coal was first mined only for local use. The remoteness and hilly geography made transportation to market costly. Not until after the Civil War and the arrival of railroads, could coal be economically mined. West Virginia is loaded with coal. Fifty-three of the fifty-five counties have coal and forty-three of these counties have coal in quantities sufficient to mine. This state leads the nation in underground coal production and in coal exports. Underground mining carries great risk.

Dangers were everywhere in the mines of the 1800s and early 1900s. Cave-ins, explosions, fires, deadly gasses and slag falls all killed and crippled thousands. Those who survived these terrors often died of black lung, a condition where coal dust eventually overcomes the lungs making it impossible to breathe, much like emphysema. It has been suggested that a U.S. soldier during WWI had a better chance of survival than a coal miner in West Virginia.1

Miners worked in muddy tunnels not large enough to stand in, with water leaking from the tunnel ceiling. Most miners worked an entire eleven-hour shift without ever standing erect, all the while breathing coal dust and hoping there was no deadly methane gas in their tunnel.

Before commercial coal mining most West Virginians made their living farming. The coal mines brought a huge migration of people. Freed black slaves moved north from the southern states and Europe's poorest immigrants flocked to the mines seeking an improvement in their lot. The coal companies set up an organization very much like the feudal system. Miners rented shabby houses without indoor plumbing from the company. Miners had to rent their tools as well and purchase their own blasting powder at inflated prices. At months end they were paid in script, which was company money good only at the company store. Miners with families often found that after they paid the company's inflated store prices they could not make ends meet. Most fell into debt to the company and received an empty pay envelope month after month.

Miners were paid by the ton of coal they extracted, 48 cents per ton in 1913. Even here the companies cheated their workers by rigging the cars to hold 2,500 ton while paying for only 2,000 pounds and docking miners for any slag or waste rock that may or may not have been in the load.

Mining conditions have gradually improved due in large part to the eventual victories of organized labor and government safety regulations. Miners still risk their lives to bring coal to the surface; ten more miners died in the years between 1997 and 2001. This is a good deal better than the 152 miners who died in the year 1968.

There are some small rewards for miners. Rare fossilized plants are occasionally found beneath layers of coal. The shale associated with coal is also the place to uncover pyrite suns. These pyrite crystals grow in a starburst disc resembling a shining sun deep in the earth. Miners often carry out these treasures in their lunch buckets to sell to geology enthusiasts.

To learn more about coal mining visit the West Virginia State Museum, Marshall University's Geology Museum or the Coal House, which is one of the few known structures made from coal.

1West Virginia's Mine Wars, compiled by the West Virginia State Archives.
Website: http://www.wvculture.org/history/minewars.html

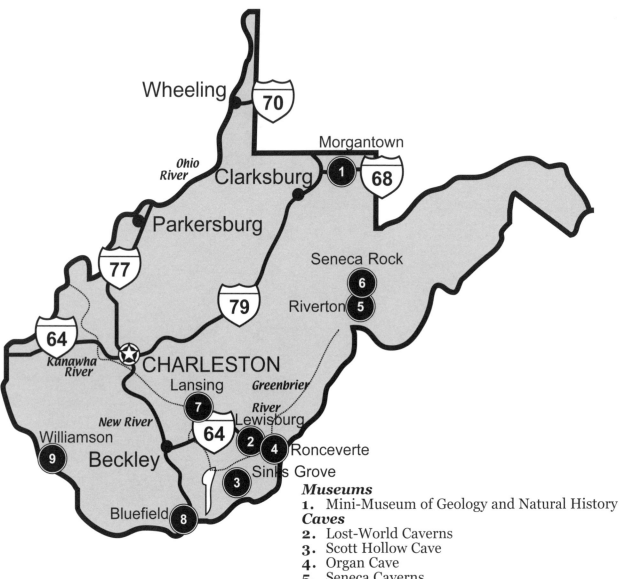

Museums
1. Mini-Museum of Geology and Natural History

Caves
2. Lost-World Caverns
3. Scott Hollow Cave
4. Organ Cave
5. Seneca Caverns
6. Smoke Hole Caverns

Points of Interest
7. Canyon Rim Visitor Center
8. Eastern Regional Coal Archives
9. Coal House

MUSEUMS

Site 1.
Mini-Museum of Geology & Natural History
Mont Chateau Research Center
P.O. Box 879
Morgantown, WV 26507-0879
(800) 984-3956
http://www.wvgs.wvnet.edu/www/museum/museum.htm
This museum showcases the rich and varied geologic history of West Virginia with displays of rock, mineral and fossil specimens.

CAVES

Site 2.
Lost-World Caverns
Route 6 Box 308
Lewisburg, WV 24901
(304) 645-6677
http://www.wvweb.com/www/lost_world_caverns/index.html
Visitors descend 100 feet to a vast chamber, decorated with flowstone formations. The mile of interconnected chambers reach depths of 235 feet below the surface. Cave attractions include, Goliath – a column forty feet tall and twenty-five feet around. The winter snowmelt caused the Crystal Waterfall. One of the world's largest compound stalactites, weighing 30 tons, creates the Snowy Chandelier. Don't forget the War Club, a stalagmite of epic proportions.

Site 3.
Scott Hollow Cave
Route 1, Box 305A
Sinks Grove, WV 24976
(304) 772-5218
http://www.wildcaving.com/
This cave was only discovered in 1984, making it a pristine example of magnificent chambers. Over the last fourteen years twenty-five miles of passages were explored. The discovery continues today. This cave makes serious effort to protect the unique underground environment. Tourists are contained to two interesting miles of cave including a half-mile of river, thus protecting the cave and its inhabitants, which include bats, salamanders and crawdads.

Site 4.
Organ Cave
417 Masters Road
Ronceverte WV 24970
(304) 645-7600
http://www.organcave.com/
Organ cave is one of the longest caves in America stretching over 40 charted miles so far. Known by pioneers as early as 1704, this site has been a popular destination for three centuries and probably much longer. The remains of a prehistoric three-toed sloth was discovered in the 1700s. In the 1800s miners dug for saltpeter for use in the American Revolution, Civil War and War of 1812. But Organ cave got its name from the unusual formation of calcite, which resembled a pipe organ.

Site 5.
Seneca Caverns
HC 78 Box 85
Riverton, WV 26814
(304) 567-2691
www.senecacaverns.com
Seneca Caverns is West Virginia's largest caverns and is located in the eastern panhandle of West Virginia, in the Appalachian Mountain range. Discovered in the 1760s, public tours did not begin until 1930. The cavern includes huge chambers of interesting formations.

Site 6.
Smoke Hole Caverns
HC 59 Box 39
Seneca Rock, WV 26884
(304) 257-4442
http://www.smokehole.com/
This cave includes a visit to Crystal Cave Coral Pool, stocked with rainbow trout, the Sparkling Room a chamber with a ceiling soaring 274 feet and loaded with stalactites and the world's largest Ribbon Stalactite, weighing six tons.

POINTS OF INTEREST

Site 7.
Canyon Rim Visitor Center
National Parks Service
U.S. Highway 19
Lansing Road
Lansing, WV 25862-0202
(304) 574-2115
http://www.nps.gov/neri/crvc.htm
Coal mining, logging and the river made this area. This visitor center contains information of natural, cultural and historic significance of the New River Gorge National River, a historic coal mining area. A video on the formation for the river gorge can be viewed here before viewing the actual gorge or hiking to the river below. Across the gorge stretches the world's longest steel-arch span bridge.

Site 8.
Eastern Regional Coal Archives
Craft Memorial Library
600 Commerce Street
Bluefield, WV 24701
(304) 325-3943
http://www.ohwy.com/wv/e/esrecoal.htm
The Coal Archives are located in the Craft Library and contain a collection of coal mining artifacts, diaries, photography, films and equipment.

Site 9.
Coal House
Tug Valley Chamber of Commerce
2nd Avenue and Court Street
Williamson, WV 25621
(304) 235-5240
http://www.coalfieldcvb.com/MingoCo/Coalhouse.html
The Coal House is a unique structure with outer walls constructed of 65-tons of coal from a multi-million dollar seam. Built in 1933 to honor this towns mining heritage, the blocks of coal were cut like stone and set upon a concrete foundation. The building currently houses the Tug Valley Chamber of Commerce. Weatherproof varnish protects the outer surface, much like paint on wood.

The Badger State gets its name from early miners who lived in the side of hills and thus were dubbed badgers

Mining is not new to this area. Early tribes mined copper and lead over 8,000 years ago in what is now Wisconsin. Early settlers arrived and mining began in the early 1800s. Copper, lead and zinc ore is found in the cracks within limestone. When immigrants from Cornwall arrived in the early 1800s they knew how to tunnel and the lead mining boom began.

Prospectors burrowed into hillsides for lead and lived in their tunnels, thus postponing the need to waste time building homes. Because of the miners' proclivity to burrow and reside underground they were dubbed "Badgers." The state seal and flag display a miner as represented by a badger, mining tools and a stack of lead. The University of Wisconsin adopted "Bucky" the badger as their mascot. The boom continued until the cry of "gold" was heard from California in 1849 and badgers headed west as fast as they could scurry, leaving their abandoned badger holes behind.

Lead strikes made men rich, they also caused environmental damage as miners stripped the forests of trees to prop up mine tunnels. Zinc and iron were discovered and mined. During WWI, 5,000 miners produced 64,000 tons of zinc.

Another important product of Wisconsin is granite. This stone composed of feldspar and quartz can withstand tremendous pressure and load, making it ideal for the construction of bridges and buildings.

Sand, gravel, lime and sandstone are all quarried in large quantity here. The sandstone is important for another reason. Prehistoric fossils are sometimes found here. Not dinosaur bones, but something much more rare and more unique - impressions of creatures without skeletons. As expected, these animals often left no trace. Therefore, there are great gaps in scientists' understanding of certain time periods due largely to the lack of evidence left behind. But here in the sandstone are found fossilized jellyfish the size of dinner plates along with other sea creatures, leaving traces of a little known time 520 million years ago.

Iron is mined in this state and used in products ranging from paperclips to skyscrapers.

Zinc is used to prevent rust and often is used to coat other metals, as in roof gutters and water tank liners. So the next time you clean out your gutters, which do not rust, thank a badger!

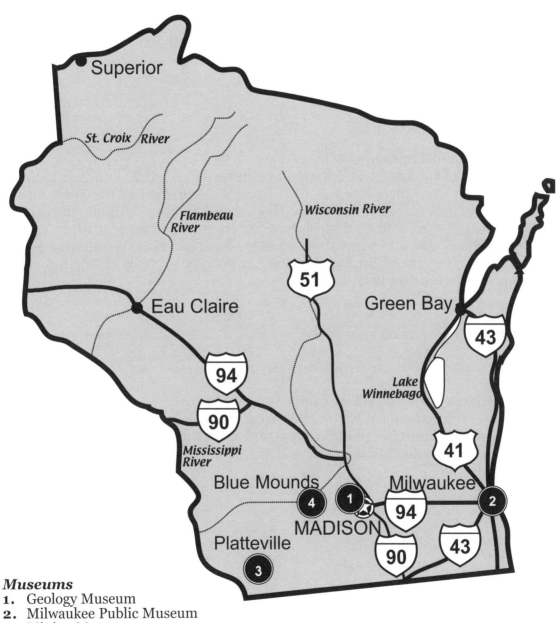

Museums
1. Geology Museum
2. Milwaukee Public Museum
3. Mining Museum

Caves
4. Cave of the Mounds

MUSEUMS

Site 1.
Geology Museum
University of Wisconsin
Department of Geology and Geophysics
1215 West Dayton Street
Madison, WI 53706
(608) 262-1412
http://www.geology.wisc.edu/~museum/
Collections of fossilized plants, marine creatures, mammals, reptiles and dinosaurs are included in this museum's collection. Many complete skeletons are exhibited as well as impressions of plants and animals.

Site 2.
Milwaukee Public Museum
800 West Wells Street
Milwaukee, WI 53233
(414) 278-2702
http://www.mpm.edu/default.asp
Two exhibits will be of special interest to fossil enthusiasts and rockhounds. The first is called "A Sense Of Wonder" and includes each of the major disciplines at the museum (including geology) in a turn of the century-style display. The second must-see exhibit entitled "The Third Planet" begins with a film introducing plate tectonics and continental drift. Visitors travel through a limestone cavern and learn about local geology, including the shallow sea that once covered Wisconsin. Then travel through the Ice Age and see animals of this time. A dinosaur diorama shows these creatures in their world.

Site 3.
Mining Museum
Main Street
Platteville, WI 53818
(608) 348-3301
http://platteville.wi.us/visitors/mining.html
The mining museum focuses on Wisconsin's historic lead and zinc mining history. A museum visit includes artifacts, dioramas, photos and a guided tour of a lead mine. Don a hardhat and descend ninety steps (fifty-feet) to the mine and get a hint of what mining was like in the 1930s.

CAVES

Site 4.
Cave of the Mounds
Brigham Farm
P.O. Box 148
Blue Mounds, WI 53517
(608) 437-3038
http://www.caveofthemounds.com
Workmen at a local limestone quarry accidentally uncovered this cave in 1939 while blasting away rock, discovering the tunnel within. The quarry was shut down after the tunnel revealed chambers and galleries filled with unique mineral formations. To protect the cave from the curious, it was closed until walkways and lights could be installed. Tourists are welcome to visit and take in the underground marvel. Above ground visit the picnic area, rock garden, visitor center and gift shop in this state park.

The following steps will help you find a likely spot on a stream or river once you have found an area in which you would like to search. A little knowledge of how gold travels in water will assist you increase your chance of finding some color.

1. First find out as much as you can about the area in which you are going to prospect. Discover where gold was found in the past, how much was extracted and if the area is continuing to produce gold today.

2. Prospectors need to learn the local regulations including where you may prospect, if dredges are allowed and if you require a permit to search.

3. If possible approach the stream or river from a high vantage point which allows you to see how the water runs, where sandbanks are, where turns occur and how fast the water flows.

4. Gold is very heavy and does not like to travel in water. So look for places where water changes from fast to slow moving. This is a spot where the gold will drop.

5. Search on the inside bends of the stream which act as a catching point for gold and other heavy materials that get moved by the force of the water.

6. Check on the downstream side of large rocks and boulders. Gold will get trapped on the rear of the rocks in little whirlpools of water and collect there.

7. Don't pass up the areas that were once covered with water during the winter run-offs. These places that are now just dry exposed rocks can contain some nice gold. These make ideal areas for highbanking, metal detecting or collecting material for a sluice box.

8. Cracks in exposed bedrock or slight gaps between large rocks make excellent collecting places for heavy material such as gold. These natural riffles can be searched with a crevice tool or screwdriver. This process is called "sniping."

9. The bottom of little waterfalls on a stream can trap gold and hold it. This is a good spot to do some dredging.

10. These are some tips that can help you start your search, but don't forget the most important tip–gold is elusive and the best place to search is where you find it!

SAPPHIRES, RUBIES, TOPAZ, DIAMOND FROM A NON-SEEDED OR DIG-YOUR-OWN SITE

These tips and techniques are intended for those who will buy buckets of non-seeded dirt or dig their own dirt directly at a mine known to contain precious gems. Methods included here involve washing your concentrates. If no water supply is available, another method of recovery must be used.

1. Be prepared to get dirty. Dress in old clothing. Some miners wear rubber boots and rain pants to keep water and mud off their duds.

2. Know that heavy material, such as diamonds, sapphires, rubies, topaz, hematite and metals sink to the bottom of all other material. This principal is vital to the methods of recovery described here.

3. It is vital that you know what you are looking for IN THE ROUGH. So take the time and allow the proprietors to show you what the diamonds, sapphires and so on, look like in their natural state. Natural stones look very different than cut and prepared stones. You are better prepared to spot it in your screen if you have seen it before.

4. Know the indicators that you are in the right place. Diamonds are only found in kimberlite. Do you know what that looks like? If not find out. Gold and Montana sapphires are often found with hematite and black sands. The pink sapphires of North Carolina are often surrounded by bits of mica and a white mud called Kaolin.

5. Realize that most precious stones are surrounded by look-a-likes. For example, many first-timers mistake pyrite for gold. Diamonds are often surrounded by flashy bits of calcite. Sapphires and topaz often nestle beside worthless clear quartz. Know how to identify imposters and realize you will find many more of these than the elusive and valuable gemstones.

6. Bring the right equipment including: shovels, work gloves, heavy-duty rubber gloves, several five-gallon buckets, a washtub, boots and rain pants. In addition you will need several classifying screens including a grizzly (1/2-inch mesh), a 1/8-inch mesh and a 1/16-inch mesh. Get a piece of foam cut to fit inside these standard screens and cut a piece of plywood to fit inside the screen as well. Finally you should bring a pocketknife, tweezers and a small closing container, such as a film container, to hold your gems.

7. Dig your material or purchase pre-classified buckets.

8. Fill your washtub half-full and stack your classifying screens from largest screen mesh on top to smallest on the bottom. Scoop the dirt you collected into the top of the stacked screens.

9. Separate clay, sand and dirt from the gravel by shaking it in water of your tub. Realize that the stones you find will be larger than the smallest screen you use. All others will collect in your washtub. If you want to catch the smallest stones you will have to later pan out your waste dirt and this takes a great deal of time. Some mines keep this excess or waste dirt, as in Montana, as it contains some gold. Continue shaking until only the gravel remains in the screens.

10. Check the grizzly and the 1/8-inch classifying screen by carefully fingering through the material and then discard the gravel in both. Remember that precious stones are very rare and you will not find one very often.

11. Take the 1/16-inch classifying screen and hold it under the water with two hands. Vigorously shake the screen from side to side until all the material lines up from twelve to six o'clock in your screen. Then turn the screen a quarter turn so the material lines up from one hand to the other (three to nine o'clock) and repeat the shaking. This action makes all the heaviest material settle to the bottom center of the screen. Repeat the process at least once more. Then lift the material from the water and gently shake it vertically to settle the material and remove excess water.

12. Now add the foam cut out on top of the screen. Place the plywood on the foam, and then place your thumbs on the board and your fingers beneath the screen. In one smooth movement, flip the screen upside down and remove the screen. The gravel should sit on the foam and board with the heavy material forming a ring in the center. If you have trouble flipping the material or find no center, gather the gravel and try again. Some miners do not use foam or a board, but have a knack for flipping out the gravel like an omelet from a pan. You may see a series of gravel circles drying on a flat patch of earth.

13. Some material needs to dry before you can check it. This includes diamonds and topaz. The look-a-likes fool the eye when wet. But when dry, clear quartz and topaz are hard to mistake. Diamonds never get cloudy as calcite and no dirt sticks to them. Some stones, such as sapphires, do not require drying. They are easily plucked with tweezers from the surrounding hematite.

14. Once the material is dry, use the tip of your pocketknife to carefully sort through the center material. Place any keepers in the safety of a closed container.

15. Some miners keep the centers of all these turnings and re-run them for a final check. Go slow and be patient.

Many years ago when the forty-niners hit the streams searching for their fortunes, they carried with them their trusty metal gold pan. This pan was used to extract gold and often to cook supper. Innovative Chinese miners even chiseled riffles in one side of the pan to catch more gold and speed the extraction process. Gold panning was and is backbreaking work. Today's pans are plastic with molded riffles so you won't have to bother chiseling your own, but you will find more gold and smaller flakes. Brand new pans need to be seasoned, so take your pan to the stream and ruff it up with dirt and rocks. This will help get any oils out of your pan. Oil attaches to gold and floats it away when you are panning. Don't eat your dinner out of your pan either, plastic makes a poor conductor so you'll have to bring your camping supplies. But at least you won't be eating from the same pan you mixed mercury in to extract unseen gold from the black sands. Yup, prospectors did that, never knowing the risks of mercury hazards on your brain. It does explain all those silly dances that miners do when they find gold.

1. First find a good spot or area with a proven record for gold and dig yourself a few buckets of material.

2. Find a calm place in the river to do your test panning. Calm waters allow you to work your material and not have to fight with the river.

3. Fill your gold pan half-full with your material, including larger rocks, sticks, clay and dirt.

4. Put your pan with material into the water. Break up clay and dirt with your hands. Wash off larger rocks and discard them. Gold sticks on a rock, gets caught in clumps of clay or dirt, so wash them in your pan.

5. With your pan still under the water slowly move the pan in a circular motion; this will get the material to move around in the pan. Allowing all of the lighter unwanted material to start to move out over the lip of your pan as the heavier material settles in the gold pan, (gold and black sand).

6. With the pan still in the water, tilt the material towards the riffles and slowly rock the pan back and forth. This helps to remove the lighter unwanted material. To avoid spilling heavy material, rock the material back by leveling your pan, then repeat the swirling motion to make sure that the heavy material again drops to the bottom of the pan. Repeat this step till you only see mostly black sand in your pan.

7. When you see mostly black sand slow down and keep a look out for those flakes of gold. A plastic sucker bottle and fine tweezers help you extract any visible gold from your pan. Don't worry about picking up some sand with the sucker bottle; you can always clean it more thoroughly later. Another useful tool is a magnet. In a plastic pan the magnet can draw hematite (black sands), which are magnetic, away from your fine gold.

Tip: Shoveling material into your gold pan through a 1/4-inch to 1/2-inch classifying screen will help you remove stones and other large material more quickly. Carefully check this material before discarding it for gold nuggets or river garnets. Good luck!

After you have found a likely place to dig for gold, collect some material and take it to a stream or river. There you can run a few test pans and discover if the location shows some color. Then use a sluice and let the water do the classifying for you. All you need is a sluice box, shovel, gloves, buckets and you are ready to find some gold.

1. Find a spot on the river that has a good depth and fast moving water. Digging is usually done near the river or stream and the concentrates brought to the sluice and fed into the upstream wider mouth of the sluice. Your digging site and sluicing site will not necessarily be beside each other, but try to keep them as close as possible so you will not have to carry your concentrates very far.

2. When digging, set an inch to a 1/2-inch mesh classifying screen over your five-gallon collecting bucket to screen out larger rocks. Check and discard the stones.

3. Place the sluice with the wide opening pointing upstream. The sluice should be almost level with the water with a slight downward pitch. Use river rocks to brace your sluice box. Place one long flat rock across the top of the sluice to keep the box in place in the current. Sometimes you need to use rocks from the river to help you change the flow of the water in the river. Such a dam can channel water toward your sluice if the current is slow or away from your sluice if the water is too fast and washing away nearly all the material.

4. Now you are ready to work some of your material through the sluice box. Slowly pour some dirt into the large opening at the top part of the sluice box. It is very important to go slowly and to regulate the amount of material that runs through the box. Give the water time to work the earth through the box.

5. Remove large rocks from the riffles to help aid the flow of the water and increase the effectiveness of the sluice. Check all rocks before you discard them, so you don't toss away a nice nugget.

6. Use a shovel to remove the tailings pile that forms at the end of the sluice box to keep the current moving and prevent back-ups in the sluice.

7. After a while, check the material in the sluice box for a build up of black sand in the riffles. This indicates its time to clean out the sluice.

8. Place a large bucket under the end of your sluice box, then slowly lift the sluice up out of the water towards the bucket. This stops the flow of water and allows you to catch all loose material in the bucket.

9. Use a gold pan to pour water through the sluice as it still sits upright in the bucket. This washes more of your material into the pail.

10. With the sluice still in the bucket, unfasten the latch and lift the metal grate. Then remove and rinse the wire mesh in the bucket with the sluice still in the bucket. Let the carpet and miners moss slide into the bucket. When the metal parts of the sluice are rinsed and clean of black sands, set them aside on the bank.

11. Now clean the miners moss by turning it upside down and shaking it in the bucket of water. Finally turn the carpet upside down and tap the surface while the carpet is submerged so any gold trapped in the carpet will fall into the bucket of concentrates. Good Luck!

States often choose a mineral, gem or rock that highlights the history, landscape or commerce. A lot can be learned about a place from the choice the legislature has made. Below is a listing in alphabetical order of our nations official gems, minerals and rocks.

STATE	GEM	MINERAL	ROCK
ALABAMA	STAR BLUE QUARTZ	HEMATITE	MARBLE
ALASKA	JADE	GOLD	
ARIZONA	TURQUOISE	FIRE AGATE	PETRIFIED WOOD
ARKANSAS	DIAMOND	QUARTZ CRYSTAL	BAUXITE
CALIFORNIA	BENITOITE	GOLD	SERPENTINE
COLORADO	AQUAMARINE		
CONNECTICUT	GARNET		
DELAWARE			SILLIMANITE
FLORIDA	MOONSTONE		AGATIZED CORAL
GEORGIA	QUARTZ	STAUROLITE	
HAWAII	BLACK CORAL		
IDAHO	STAR GARNET		
ILLINOIS	FLUORITE	FLUORITE	
INDIANA			LIMESTONE
IOWA			GEODE
KANSAS			
KENTUCKY	FRESHWATER PEARL	COAL	KENTUCKY AGATE
LOUISIANA	AGATE		PETRIFIED PALM
MAINE	TOURMALINE		
MARYLAND			
MASSACHUSETTS	RHODONITE	BABINGTONITE	ROXBURY PUDDINGSTONE GRANITE
MICHIGAN	GREENSTONE (CHLORASTROLITE)		PETOSKEY STONE
MINNESOTA	LAKE SUPERIOR AGATE		
MISSISSIPPI			PETRIFIED WOOD
MISSOURI		GALENA (LEAD)	MOZARKITE
MONTANA	YOGO SAPPHIRE MOSS AGATE		
NEBRASKA	BLUE AGATE		

STATE	GEM	MINERAL	ROCK
NEVADA	PERIDOT, BLACK FIRE OPAL, TURQUOISE	SILVER	SANDSTONE
NEW HAMPSHIRE	SMOKY QUARTZ	BERYL	CONWAY GRANITE
NEW JERSEY			
NEW MEXICO	TURQUOISE		
NEW YORK	GARNET, BLACK TOURMALINE, MOONSTONE	HEMATITE	
NORTH CAROLINA	EMERALD		GRANITE
NORTH DAKOTA			TEREDO WOOD
OHIO	OHIO FLINT		
OKLAHOMA			BARITE ROSE
OREGON	SUNSTONE		THUNDEREGGS
PENNSYLVANIA			TRILOBITE
RHODE ISLAND		BOWENITE	CUMBERLANDITE
SOUTH CAROLINA	AMETHYST		BLUE GRANITE
SOUTH DAKOTA	FAIRBURN AGATE	ROSE QUARTZ	
TENNESSEE	TENNESSEE PEARL	AGATE, TENNESSEE LIMESTONE	
TEXAS	BLUE TOPAZ		PETRIFIED PALMWOOD
UTAH	TOPAZ	COPPER	COAL
VERMONT		GROSSULAR GARNET, TALC	GRANITE, MARBLE, SLATE
VIRGINIA			CHESAPECTEN JEFFERSONIUS (FOSSIL)
WASHINGTON	PETRIFIED WOOD		
WEST VIRGINIA	LITHOSTROTIONELLA (WEST VIRGINIA FOSSIL CORAL)		
WISCONSIN	RUBY	GALENA (LEAD)	RED GRANITE
WYOMING	JADE		

The association of stones to the month of a person's birth began about 400 AD. But birthstones were not commonly worn by those born in each month until the 1700s. There is some evidence that stones were not worn according to the month of your birth, but a set of all twelve stones might be obtained and worn in sequence according to the date in order to maintain one's health and for personal protection. If you consider this nothing but superstition and the working of an over-active imagination, consider how important a person's state of mind is to their health and well-being. The stones on the list below have changed only slightly through the ages with some remaining unchanged by the year.

JANUARY

Modern–Garnet
Ancient–Garnet
"No gems save garnets should be worn
By her who in this month is born;
They will insure her constancy,
True friendship and fidelity."[1]

The word garnet is derived from the Latin word for grain and a Middle English and old French word for pomegranate, a fruit known for its deep red fruit. Because of their color, garnets were used in ancient time to cure conditions of the blood and are symbols of love and passion. Some garnets are such a deep red they are mistaken for rubies, although red is not the only color in which garnet occur. Garnets from Ceylon range from a cinnamon-yellow to reddish-orange. The most rare and valuable of all is the grossularite garnet known as the green garnet.

FEBRUARY

Modern–Amethyst
Ancient–Amethyst
"The February-born may find
Sincerity and peace of mind,
Freedom from passion and from care,
If she an amethyst will wear."[1]

The word amethyst comes from the Greek and roughly means "one who is not drunk." It was thought that water drunk from a goblet of purple glass resembled wine. Those who drank from such a cup did not get drunk and therefore amethyst was believed to protect from drunkenness. During the Renaissance period, amethyst adorned religious objects including Bishop's rings. Amethysts are the most popular of the quartz family and range in color from red-violet to blue-violet and true purple.

MARCH

Modern–Aquamarine
Ancient–Jasper
"Who on this world of ours her eyes
In March first opens may be wise,
In days of peril firm and brave,
Wears she a bloodstone to her grave."[1]

Aquamarine derives its name from its resemblance to sea water. This stone is popular among many sailors who believed it protected them from the perils of the sea. Egyptians wore this stone in battle to protect and instill courage. In the Middle Ages, aquamarine earned another use, the guarantee of happy marriages and fidelity in newlyweds. The color of this blue beryl ranges from light blue to blue-green to dark blue.

APRIL

Modern–Diamond
Ancient–Diamond
"She who from April dates her years
Diamonds would wear, lest bitters
For vain repentance flow. This stone
Emblem of innocence is known."[2]

Diamonds, pure crystallized carbon, have become the most prized of gemstones. This gem wins the prize for hardness and comes in an astonishing variety of colors. Make sure to accept only the best. At one time, Hindus believed that wearing a flawed diamond could keep them from the highest level of heaven. Diamonds were used as antidote for poison, though some believed the stone itself was poison, as it was said to be found in places guarded by a vile venomous beast that imbued its treasure with poison. It was noted in the Middle Ages that poor people more often succumbed to plague and the possession of diamonds by the rich became the assumed reason. Diamonds were said to protect from plague and petulance, cure numerous diseases, including those of the bladder. How romantic.

MAY

Modern–Emerald
Ancient–Emerald
"Who first behold the light of day
In spring's sweet flow'ry month of May,
And wears an emerald all her life,
Shall be a loved and happy wife."[2]

The name for emeralds is Greek and means green stone. Emerald is believed to improve intelligence and cure diseases of the mind and heart. That makes it perfect to protect marriages, increase fertility and protect in childbirth. This stone is softer and more fragile than sapphires, rubies and diamonds. Jewelers often cut it in the strong emerald cut to protect it from damage. Emeralds vary in hue from light to dark and often include inclusions and black specks of carbon.

JUNE

Modern–Alexandrite
Ancient–Agate
"Who comes with summer to this earth,
And owes to June her hour of birth,
With ring of agate on her hand
Can health, long life and wealth command."[3]

Alexandrite is a newcomer to the gem market arriving in 1830. The stone derives its name from Prince Alexander as it was discovered on the day he came of age. This gem is unique as it changes in color. Light or dark green in sunlight it changes to red in candlelight or tungsten light. Most valued is brilliant green changing to fiery red. As it is a modern stone, there is no folklore or ancient superstition associated with this stone.

July

Modern–Ruby
Ancient–Turquoise
"The heav'n-blue turquoise should adorn
All those who in July are born;
For those they'll be exempt and free
From love's doubts and anxiety."[3]

From the Latin for red, ruby is symbolic of love, charity and victory. In ancient times, it was believed to sooth anger and bring courage. The red color caused early people to associate rubies with blood and fire. They were symbolic of war, battle, victory and therefore adorned by royalty, valued above diamonds. It was not until modern times that it was discovered that rubies have a twin. Sapphires are identical to rubies, both being corundums and differing only in color. Sapphires come in numerous colors, while rubies are always and only red, varying only in shade.

August

Modern–Peridot
Ancient –Carnelian
"Wear a carnelian or for thee
No conjugal felicity;
The August-born without this stone,
'Tis' said, must live unloved, alone."[3]

Peridot is a variety of olivine. Peridot is purported to bring the wearer peace, luck and success. Peridot has never enjoyed the popularity of the big three, diamond, sapphire and ruby. But dark peridot may even have slipped into queen Cleopatra's emerald jewelry. Peridot was believed in ancient times to derive its power from the sun and thought to have strong medicinal powers. Peridot was believed to protect against enchantments, the evil eye and nightmares. Perhaps because of its soothing color, this stone was also believed to attract love and calm anger. Among its more mundane uses is the healing of insect bites. Peridot varies in color from bright yellow-green to green.

September

Modern–Sapphire
Ancient–Chrysolite (Peridot)
"A maid born when September leaves
Are rustling in the autumn breeze,
A chrysolite on brow should bind–
'Twill cure disease of the mind."[4]

The name sapphire comes from the Greek word for blue. This corundum actually comes in all varieties of color except red, which is reserved for rubies. Any other color of corundum is considered a sapphire. This stone has the celestial beauty of the heavens and has always been highly prized. It has been and still is considered a token of lasting love. Sapphires symbolize truth, tradition and sincerity. One thing is certain; they are tough, falling second only to diamond in hardness.

October

Modern–Rose Zircon, Opal
Ancient–Beryl
"October's child is born of woe,
And life's vicissitudes must know;
But lay a beryl on her breast,
And Hope will lull those woes to rest."[4]

The name may have come from the Roman word, opalus and the Greek word, opallios, meaning "to see a change of color." Of all the gemstones, opal is most unusual and has been described as having the fire of emeralds, sapphires and rubies. The amazing variety of colors in opals makes them highly prized. Opals were thought to improve eyesight and keep hair from turning white. This gem's most spectacular use is its ability to render the wearer invisible, though I wouldn't try this one at home. Opal is not a crystal but a dehydrated silica jelly that cracks and thereby reflects light. The most valuable opals are dark gray or black. Since opals contain some water they are more fragile than other gemstones.

November

Modern–Topaz
Ancient–Topaz

"Who first comes to this world below
With drear November's fog and snow
Should prize the topaz's amber hue–
Emblem of friends and lovers true."[5]

The name topaz comes from Greek and is derived from the location of the mines on the Island of Topazos. Topaz was used in ancient times to cure poor vision. During the black plague, topaz was pressed to sores to heal them. This gem is renowned for its curative powers. Topaz is a silicate of aluminum and occurs in yellow, pink, blue, brown and colorless variety.

December

Modern–Blue Zircon
Ancient–Ruby

"If cold December gives you birth–
The month of snow and ice and mirth–
Place on your hand a ruby true;
Success with bless whate'er you do."[5]

The word zircon comes from Persian language meaning golden, as was its original color of zircon when discovered. Blue zircon is a recent addition to the world of gemstones and became popular in the 1920s. For many years it was considered to possess mystic properties because it was noticed that stones occasionally changed color. Zircon is known to occasionally form in association with uranium and thorium, which over time could account for the color change. This magic transformation must have been quite disconcerting in ancient times. The most prize color is an electric blue. It is also occurs in pale blue, sky blue, greenish-blue. Zircon is often heated to enhance its natural color.

Footnote
***

[1] Kunz, *The Curious Lore of Precious Stones*, 1971, p.327
[2] Kunz, *The Curious Lore of Precious Stones*, 1971, p.328
[3] Kunz, *The Curious Lore of Precious Stones*, 1971, p.329
[4] Kunz, *The Curious Lore of Precious Stones*, 1971, p.330
[5] Kunz, *The Curious Lore of Precious Stones*, 1971, p.331

As you can see the marriage must survive eleven long years before the couple merits the gift of the most precious objects–gemstones.

YEAR	GIFT
1	Paper
2	Calico
3	Linen
4	Silk
5	Wood
6	Candy
7	Floral
8	Leather
9	Straw
10	Tin
12	Agate
13	Moonstone*
14	Moss-Agate
15	Rock Crystal/Glass
16	Topaz
17	Amethyst
18	Garnet
19	Rose Quartz
20	Jade
23	Sapphire
25	Silver
26	Star Sapphire, blue*
30	Pearl
35	Coral/Jade
39	Cats-eye*
40	Ruby
45	Alexandrite
50	Gold
52	Star Ruby*
55	Emerald
60	Diamond, yellow
65	Star Sapphire, gray*
75	Diamond

*All anniversaries, which are multiples of 13, have gems believed to counteract the bad influence of this unlucky number.[1]

Footnote

---

[1]Kunz, *The Curious Lore of Precious Stones*, 1971, p.337

The biggest and brightest gems and nuggets capture our imagination. Some of these are old finds and come with long and strange histories. Some were found only in the last few years. Fantastic discoveries are made each year. But you have to get out there to make one. All of the minerals, gems and metals listed below are still being recovered. Best of all, there is a site somewhere in the U.S. open to the public that allows you the opportunity to find them. This listing is by no means a complete one. How could it be when new discoveries are being made? It is merely presented as a way to inspire you to your own discoveries.

GOLD

- The largest mass of gold ever found is the Holtermann Nugget. This monster nugget was discovered on October 19, 1872 in the Beyers & Holtermann Star of Hope Mine in Australia. It weighed 517 pounds and yielded 180 pounds of gold.

- The largest solid gold nugget is called "the Welcome Stranger" and was discovered in Australia in 1869 in a wagon rut. The lustrous metal was as heavy as the mule driver who found the 210-pound nugget.

- The largest nugget found in America was unearthed in California during the Gold Rush. Found in the Morgan Mine in Carson Hill on November of 1854, it weighed an incredible 214 pounds.

- Alaska's biggest nugget proves that the best may still be out there. "The Alaska Centennial Nugget" was discovered in 1998 and weighed 24 pounds (294 Troy ounces). A miner in Ruby, Alaska noticed the massive nugget rolling before the bulldozer blade.

- A child of twelve in 1799 spotted the largest gold nugget found on the East Coast. The boy, Conrad Reed, found the 17-pound nugget in a local stream in Cabarrus County, North Carolina. He brought it home and for three years it served as a family doorstop. A local jeweler paid his father a week's wages for the gold, $3.50. This was less than one percent of the gold's value at the time.

- Colorado was home to a nugget weighing over 17 pounds. Discovered in 1887 by Tom Groves, the miner brought his precious find to the assayer's home wrapped in a blanket. "Tom's Baby," as it became known, can be seen at the Denver Museum of Nature & Science in Colorado.

- Georgia is also home to a number of large gold nuggets, many weighing over 2 pounds.

SILVER

- The largest single silver nugget, "the Smuggler Mine Nugget," was found in 1893 in a mine near Aspen, Colorado and weighed 2,350 pounds. Ironically, it was not removed from the mine until 1894 because silver prices crashed the year of its discovery.

- In the same year, the Molly Gibson Mine yielded a silver nugget weighing a massive 1,840 pounds. Ironically, both of the largest nuggets ever found were discovered at the worst possible time, the year of the crash of the silver market.

PLATINUM

- The largest known platinum nugget was found in the Ural Mountains in Russia in 1843. This nugget weighed over 21 pounds and was melted down almost immediately after its discovery.

- The second largest platinum nugget was also found in the Ural Mountains and tipped the scales at over 17 pounds. This one survived the melting pot.

- Here in the U.S. you can see platinum at most major museums. The Natural History Museum of Los Angeles has a 0.5 Troy ounce nugget recovered in a stream in California.

DIAMONDS

- The largest diamond known in existence is "the Cullinan," discovered in January of 1905 in South Africa and weighing 3,106 carats. This massive diamond yielded 106 cut stones including one weighing 530.20 carats. This stone called "the Star of Africa" was cut in a pear shape with 74 facets. It now graces the Royal Scepter of England and is kept with the Crown Jewels in the Tower of London.

- The largest cut diamond is the "Golden Jubilee" diamond weighing 545.67 carats. It is currently set in the Thai Royal Scepter.

- The largest uncut diamond is owned by De Beers, of course, and weighs 1,462 carats.

- The largest mounted diamond currently resides in the Smithsonian Institute weighs 127.01 carats and is called "the Portuguese Diamond." The owner, Mr. Harry Winston traded it for 2,400 carats of small diamonds in 1963. The gem is a flawless and octagonal cut white diamond. The history of this stone is sketchy. This is puzzling and efforts have been unsuccessful to discover its origins.

- The most unique diamond is the "Hope." The lovely blue diamond of 45.52 carats is believed by some to be cursed. It certainly was not lucky for a previous owner and former Queen of France, Marie-Antoinette, who was beheaded.

- The largest diamond discovered in North America is "the Uncle Sam," a 40.23 carat white diamond unearthed in a stream in Arkansas in 1924. This diamond was later emerald cut to a weight of 12.43 carats. This American treasure is on display at the Museum of Natural History in New York City.

- Smaller specimens are on display in most large museums. The Natural History Museum of Los Angeles has a 1.95 carats uncut diamond recovered from the Smithflat area in 1896. The Crater of Diamonds State Park in Arkansas has a small museum that features many uncut diamonds. The museum also has castings of the biggest gems recovered from this area, including the "Uncle Sam."

RUBIES

- One of the world's most famous rubies is the "Timur Ruby," weighing 352.50 carats. It was thought to be the largest ruby in existence but it is not a ruby at all, but a rare red spinal. The spinal is part of a necklace belonging to Elizabeth II, the Queen of England, and worn on state occasions.

- The largest star ruby is the "Eminent Star Ruby," an oval cabochon with six-ray star. This stone has a carat weight of 6,465 and is believed to come from India.

- The "Burma Hixon Ruby" was unearthed in one of the most renowned mines at Mogok in India. It weighs 196 carats and is owned by the Natural History Museum of Los Angeles County.

- The Black Prince's Ruby is as large as a hen's egg and weighs approximately 170 carats. This gem found its way to England and is set in the state crown. Unfortunately, it is also not a ruby but another red spinal.

- Nearer home is a red star ruby residing at the Natural History Museum of Los Angeles. The gem is 18.29 carats and is on display in the Hixon Gem Vault exhibit. This one is really a ruby.

EMERALDS

- The largest carved emerald in the world was found in Brazil in August of 1974. It weighed 86,136 carats and was carved by an artisan named Richard Chan from Hong Kong.

- Another famous emerald is "the Hooker Emerald," once owned by the sultan of the Ottoman Empire, and reportedly was worn as a belt buckle. This gem was acquired by Tiffany & Co. and set with 901 diamonds in a tiara. This setting changed again in the 1950s to a brooch and was donated to the Smithsonian Institution in 1977. It is currently on display in Washington, DC.

- The largest emerald found in North Carolina was unearthed in the mid-1980s and weighed 858 carats. Originally named "the Jolly Green Giant," this uncut gem is now on display in the North Carolina Museum of Natural Science and renamed the more dignified "the Empress Caroline."

- The largest cut emerald found in North America also comes from Hiddenite, North Carolina. Found in 1980, the gem weighs 15.46 carats and has two names, "the Kite Emerald" or "the June Culp Zeitner Emerald."

SAPPHIRE

- The largest star sapphire is called "the Lone Star" and weighed 9,719.5 carats. The stone was cut in London in 1989.

- "The Star of Asia" is another beautiful star sapphire on display in the Smithsonian Institute. This 330 carats gem is royal blue in color.

- "The Logan Sapphire" is a faceted gem of 423 carats. This gem was mined in Sri Lanka and is roughly the size of a chicken's egg.

- An uncut sapphire that was 3,965 carats from Sri Lanka may be the largest uncut specimen known in existence. It is the size of a fist and is expected to yield somewhere between fifty and one hundred cut gems.

OPAL

- The largest opal is a single piece of white opal weighing 26,350 carats. Found in 1989 in Australia it is called "the Jupiter Five."

- The largest known opal bearing rock is the 7-ton "Painted Lady," found in the Andamooka Opal Fields in South Australia. This monster rock is over nine feet long and four feet wide and contains veins of opal up to an inch thick.

- The largest black opal was also found in Australia in 1972, and produced a finished gem weighing 1,520 carats called "the Empress of Glengarry."

JADE

- Largest piece of Jade was found in Canada in 1992 and weighed 577 tons or 1,154,000 pounds.

- The Emerald Buddha is a 30-inch statue carved of jadeite and housed in its own sacred temple in Bangkok, Thailand. The statue was discovered in 1434 when lightning struck a pagoda and revealed an unimpressive stucco Buddha. Over time the plaster fell away and the beautiful jade Buddha was revealed. The Emerald Buddha is thought to bring good fortune and several miracles are attributed to it. Want more? Well, this little Buddha has a golden wardrobe that changes with the season. In winter he wears a golden shawl and when it rains he has a golden hat and above his head are seven golden umbrellas.

Amber

- The largest piece of amber weighed in at over 33 pounds. It was discovered in Burma and now resides in the Natural History Museum in London.

Topaz

- Two of the largest topaz crystals ever unearthed were discovered in Brazil and now reside in the Smithsonian Institution in Washington, DC. They weigh 111 pounds and 70 pounds respectively and still exist in their natural state because of the discovery nearby of a large 10-pound topaz crystal more suitable for cutting. The result is the "American Golden Topaz" which has a carat weight of 22,892.5 and contains 172 facets. This rare gem is nearly 17-inches tall and over 14 inches wide.

Fossils

- Do you know Sue? Sue is the largest Tyrannosaurus Rex skeleton ever recovered. There are only twenty-two such skeletons known to exist and she's the biggest. This fossil is named for her discoverer, Sue Hendrickson, who found it in 1990 in the badlands of South Dakota. Sue is an extraordinary ninety percent complete and now lives in Chicago in The Field Museum.

- The Carcharodon Megalodon was a prehistoric ancestor of the great white shark. This swimming sea monster was as large as a city bus and shed hundreds, perhaps thousands of teeth in a lifetime. These teeth are recovered today by divers in rivers in South Carolina and off the coast of Venice, Florida. Teeth over five-inches long are rare but not unheard of.

Shipwreck Treasure

- The most spectacular shipwreck recovery is the wreck of the *Central America*. This ship sank in a hurricane in 1857 off the coast of the Carolinas in very deep water (eight thousand feet). The *Central America* was a side-wheel steamer carrying a bounty of the California gold fields. The ingots, nuggets and gold dust and freshly minted coins measured many tons. Tommy Thompson discovered the wreck in 1989. This project was the first ship ever to be salvaged from such a depth and used robots to recover the treasure.

- A wreck of spectacular wealth salvaged in shallow water is the *Nuestra Senora de Atocha* found in fifty feet of water by Mel Fisher after a staggering sixteen-year search. Somewhere in the warm water off Key West this ship sank in a hurricane in 1622. As one of the Spanish treasure ships, it was laden with New World wealth. The recovered treasure is estimated to be worth over four hundred million. Some of the booty is on display in Key West at the Mel Fisher Maritime Heritage Museum, including a twelve-foot gold chain weighing six pounds and a bishop's cross set with seven magnificent emeralds from Columbia. The search continues today for the back half of the ship.

- If fifty feet is still too deep, how about the 1715 fleet? This armada of twelve ships sank off the barbarous coast of Florida, now Vero Beach, in a hurricane. In 1997, a gold butterfly brooch set with 161 diamonds, a second brooch set with 170 diamonds and two earrings each containing 53 diamonds were found only seventy-five feet from the beach. The jewelry is estimated at $1.2 million. Although water searching is prohibited by salvage claims, beach hunting is permitted. A lucky metal detector enthusiast recovered a gold whistle in the shape of a dragon on the beach after a storm.

Admiralty Claim: A specific type of claim leased to salvaging companies to guarantee their exclusive use of a certain area of ocean bottom. This means you may not use a metal detector in the area of the wreck sites.

Back Hoe: Large machine similar to a bulldozer, which moves dirt and uncovers new material to mine.

Black Sand: Black grains of sand made of hematite. This is a very heavy metal that is often associated with gold. Seeing it in your gold pan is a good sign.

Bureau of Land Management (BLM): National organization created to protect, preserve and manage public land. Some mining and rock collecting is permitted on BLM land in designated areas. Permits are often required.

Carat: The unit used to express the weight of gemstones. One carat is equal to two hundred milligrams.

Claims: A tract of land staked out by an individual to mine. This means it is not open to you and is protected by law.

Concentrates: Dirt that has been reduced by removing all large rocks, light soil and organic material. This leaves just concentrated, heavy dirt that is likely to contain heavy gold or gemstones.

Discriminator: A function of a metal detector, used to tune out junk metal such as flip tops and tin foil. It is also used to tune out mineralized soil or rocks (hot rocks).

Dredge: Mining equipment used in a stream to find gold. It works like a powerful water vacuum on the streambed. It sucks up the bottom soil. Heavy material including gold falls onto a ridged mat, while light material washes through and falls back to the stream floor.

Dump: Where the non-valuable rock and debris is discarded at a mine site.

Escudo: Spanish coin made of gold. An escudo weighs one ounce. They are also commonly called a doubloon. They were minted in South America by the Spanish conquistadors from the 1600s to the 1700s. Gold was then shipped to Spain. Some treasure galleons were lost in storms along the U.S. coast.

Gold Dust: Fine particles of gold found in rivers, streams, oceans, soil, and hopefully, your gold pan. Finding gold dust indicates you have found a good place to search more thoroughly.

Hand-Operated Tools: Tools used by hand. Some examples include shovels, picks, garden trowels, garden rakes and pry bars.

Highbanker: This is a desert sluice box. It is used in places where water is unavailable or inconvenient. It has its own supply of recycled water that wets dirt and sends it into a sluice box. Some highbankers use hot air instead of water.

Hot Rocks: Rocks with a concentration of metal or salt that cause metal detectors to give a false signal. This signal leads metal detectors to think they have found something of value. Hot rocks should be checked. If quartz is found you may have hope that gold may still be present.

Mesh Screens: Mesh screens are used to remove larger rocks and organic material from dirt in an effort to concentrate it. Mesh screens are commercially produced and easily fit into five-gallon buckets. Many miners make their own with chicken wire and wood. Sizes describe the size of the mesh openings. The most common sizes are 1/4-inch and 1/2-inch mesh screens.

Mine Run: A large quantity of mined material offered for sale. It may include as much material as is produced in a half day or full day of mining. The quantity of material included varies from mine to mine, so get specifics before you buy.

Mineralization: Soil that has a concentration of minerals, metals or salts. This plays havoc with metal detectors which read it as valuable material such as gold or silver. Rocks may also be mineralized and are called hot rocks. Wet ocean sand is highly mineralized. A metal detector with a good discriminator is needed in cases of mineralized material.

Mini Jig: A mining tool that washes dirt and removes light waste rock and organic material. The condensed material containing heavier metal is collected for more careful inspection. It is used to find such materials as garnets, gold, rubies and sapphires.

Native Rock: The rock most prevalent to a specific area. The rock surrounding or being penetrated by mineral veins and deposits.

Placer Gold: Gold found in its natural state without any host rock. Examples include gold dust and gold nuggets. Placer gold is commonly found in streams and rivers where it has washed away from its source.

Pry Bar: A tool made of steel that is used to move a heavy rock or to turn over a rock to expose material underneath. A crow bar is a large pry bar. They come in a variety of sizes.

Reale: A Spanish coin made of silver. A one-ounce coin was called an eight reale. This is where the term "pieces of eight" came from. They were minted in South America by the Spanish from the 1600s to the 1700s.

Seeded: This is dirt offered to you by a mine that has had objects added to it to ensure that you are not disappointed. Material added is not always indigenous to the area. Gemstones added are usually of inferior quality.

Sluice Box: This piece of mining equipment is used in combination with a running stream. The water runs through one end of the box and out the other. Material from the streambed is added to the top of the box. Light material runs off with the water, while the gold and heavy materials fall to the mat in the sluice box for closer examination.

Snuffer Bottle: A plastic bottle used to suck up fine gold dust from your gold pan.

Tailings: Term referring to mining material that has been discarded during the mining process, also called waste rock. This is not usually the best material to search, but can contain fine specimens and is much easier than digging your own.

Ultraviolet Light: A type of light bulb that is long or short wave. It is also called a black light. This light is used to spot rocks that fluoresce, or glow in a variety of colors as the ultraviolet light is shone upon them.

Under Claim: Area of land that has been claimed by an individual or company to mine. This means it is not open to you and is protected by law. Claim markers mean keep off and should be respected.

Annual Buyers' Guide, *Lapidary Journal*. Boulder: Primedia, Inc., 2000.

Bauer, Jaroslav and Vladimir Vouska. *A Guide in Color to Precious & Semiprecious Stones*. Chartwell Books, Inc., 1992.

Cipriani, Curzio. *Simon & Schuster's Guide to Gems and Precious Stones*. New York: Simon & Schuster, 1986.

Cunningham, Scott. *Cunningham's Encyclopedia of Crystal, Gem and Metal Magic*. St. Paul: Llewellyn's Publications, 1995.

Cross, Brad. *Gem Trails of Texas*. Baldwin Park: Gem Guides Book Co., 2001

Gold Prospectors Association of America. *1994 Gold Prospector's Mining Guide*. Temecula: Gold Prospectors Association of America, 1994.

Kunz, George Frederick. *The Curious Lore of Precious Stones*. New York: Dover Publications, 1971.

Lawless, Chuck. *The Old West Sourcebook, A Traveler's Guide*. Crown Publishers, 1994.

Mitchell, James R. *Gem Trails of Arizona*. Baldwin Park: Gem Guides Book Co., 2001.

Mitchell, James R. *Gem Trails of Colorado*. Baldwin Park: Gem Guides Book Co., 1997.

Mitchell, James R. *Gem Trails of Nevada*. Baldwin Park: Gem Guides Book Co., 2002.

Mitchell, James R. *Gem Trails of New Mexico*. Baldwin Park: Gem Guides Book Co., 2001.

Mitchell, James R. *Gem Trails of Northern California*. Baldwin Park: Gem Guides Book Co., 1995.

Mitchell, James R. *Gem Trails of Southern California*. Baldwin Park: Gem Guides Book Co., 1996.

Moore, Barry. Herkimer Diamonds, *A Complete Guide for the Prospector and Collector*. New York: Herkimer Diamond Development Corp., 1989.

Petralia, Joseph, F. *GOLD!, GOLD!, A Beginner's Handbook and Recreational Guide*. San Francisco: Sierra Trading Post, 1999.

Post, Jeffrey E. *The National Gem Collection*. New York: Harry Abrams, Inc., 1997.

Reader's Digest: Off the Beaten Path. Reader's Digest Inc., 1997.

Reilly, Kevin; Rowe, Gary T; and Marnville, Kevin. *Hurricane Treasure, 1715 Beach Sites, Locations Revealed*. Pompano Beach: Pirate Express Publishing, 1990.

Symes, R.F. and R.R. Harding. *DK Smithsonian Eyewitness Book: Crystal and Gem*. London: Dorling Kindersley, 1991.

Ward, Fred. *Diamonds*. Bethesda: Gem Book Publishers, 1993.

Ward, Fred. *Emeralds*. Bethesda: Gem Book Publishers, 1993.

Ward, Fred. *Rubies & Sapphires*. Bethesda: Gem Book Publishers, 1995.

White, John Sampson. *The Smithsonian Treasury Minerals and Gems*. Washington, DC: Smithsonian Institution Press, 1991.

http://www.ucmp.berkeley.edu/subway/nathistmus.html, The University of California Museum of Paleontology, Berkeley & the Regents of the University of California, 2002.

http://www.goodearthgraphics.com/showcave.html, The United States Show Cave Directory, 1996.

http://cavern.com/caves.htm, National Cave Association, Park City, Kentucky, 2001.

http://www.jewelrymall.com/stategems.html, Jewelry Mall, 1999.

http://sln.fi.edu/tfi/hotlists/museums.html, The Franklin Institute Online, Philadelphia, Pennsylvania, 2001.

http://www.sfmuseum.org/hist7/tencom.html, Gladys Hansen, Museum of the City of San Francisco, 2001.

MAGAZINES

The American Mineralogist
Mineralogical Society of America
1015 18th Street NW, Suite 601
Washington, DC 20006

Gems and Gemology
Gemological Institute of America
5345 Armada Drive
Carlsbad, CA 92008

Lapidary Journal
60 Chestnut Avenue, Suite 201
Devon, PA 19333-1312

Mineralogical Record
P.O. Box 35565
Tucson, AZ 85740

Rock & Gem
4880 Market Street
Ventura, CA 93003

Rocks & Minerals
Heldref Publications
1319 18th Street, NW
Washington, DC 20036-1802

BOOKS

IDENTIFYING ROCKS & MINERALS

Chesterman, Charles W. *Audubon Field Guide to North American Rocks and Minerals*. Alfred A. Knopf, Inc., New York, New York, 1979.

Fejer, Eva and Cecelia Fitzsimons. *An Instant Guide to Rocks and Minerals*. Gramercy Books, New York, New York, 1988.

Horenstein, Sidney, editor. *Simon & Schuster's Guide to Fossils*. Simon & Schuster Inc., New York, New York, 1986.

Maley, Terry S. *Field Geology Illustrated*. Mineral Land Publications, Boise, Idaho, 1994.

Pellant, Chris. *DK Smithsonian Handbook: Rocks and Minerals*. Dorling Kindersley, Ltd., London, 1992.

Pough, Frederick H. *Petersen Field Guides® Rocks and Minerals*, 5th ed. Houghton Mifflin Co., New York, New York, 1996.

Ricciuti, Edward and Margaret Carruthers. *Petersen First Guide® Rocks and Minerals*. Scholastic Inc., New York, New York, 1998.

Prinz, Martin, George Harlow, and Joseph Peters, editors. *Simon & Schuster's Guide to Rocks and Minerals*. Simon & Schuster Inc., New York, 1977.

Zim, Herbert S. and Paul R. Shaffer. *Rocks and Minerals*. St. Martin's Press, Racine, Wisconsin, 2001.

FIELD COLLECTING

Blair, Gerry. *Rockhounding Arizona*. Falcon Press, Helena, Montana, 1992.

Butler, Gail A. *Rockhounding California*. Falcon Press, Helena, Montana, 1995.

Cross, Brad. *Gem Trails of Texas*. Gem Guides Book Co., Baldwin Park, California, 2001.

Crow, Melinda. *Rockhounding Texas*. Falcon Press, Helena, Montana, 1994.

Ettinger, L. J. *Rockhound and Prospector's Bible*, 3rd ed. L. J. Ettinger, Reno, Nevada, 1992.

Feldman, Robert. *Rockhounding Montana*. Falcon Press, Helena, Montana, 1985.

Girard, Roselle M. *Texas Rocks and Minerals: An Amateur's Guide*, rev. ed. Bureau of Economic Geology, University of Texas, Austin, Texas, 1964.

Johnson, H. Cyril and Robert N. Johnson. *Coast to Coast Gem Atlas*, 5th ed. Cy Johnson & Son, Susanville, California, 1987.

Kelty, David. *The GPS Guide to Western Gem Trails*. Gem Guides Book Co., 2002

Kimbler, Frank S., and Robert J. Narsavage, Jr. *New Mexico Rocks and Minerals*. Sunstone Press, Santa Fe, New Mexico, 1981.

Krause, Barry. *Mineral Collector's Handbook*. Sterling Publishing Co., Inc., New York, New York, 1996.

Mitchell, James R. *Gem Trails of Arizona*. Gem Guides Book Co., Baldwin Park, California, 2001.

Mitchell, James R. *Gem Trails of Colorado*. Gem Guides Book Co., Baldwin Park, California, 1997.

Mitchell, James R. *Gem Trails of Nevada*. Gem Guides Book Co., Baldwin Park, California, 2002.

Mitchell, James R. *Gem Trails of New Mexico*. Gem Guides Book Co., Baldwin Park, California, 2001.

Mitchell, James R. *Gem Trails of Northern California*. Gem Guides Book Co., Baldwin Park, California, 1995.

Mitchell, James R. *Gem Trails of Oregon*. Gem Guides Book Co., Baldwin Park, California, 1997.

Mitchell, James R. *Gem Trails of Southern California*. Gem Guides Book Co., Baldwin Park, California, 2003.

Mitchell, James R. *Gem Trails of Utah*. Gem Guides Book Co., Baldwin Park, California, 1996.

Mitchell, James R. *The Rockhound's Handbook*. Gem Guides Book Co., Baldwin Park, California,1997.

Ream, Lanny R. *Gems and Minerals of Washington*. Jackson Mountain Press, Renton, Washington, 1994.

Ream, Lanny R. *Idaho Minerals*. L. R. Ream Publishing, Coeur d'Alene, Idaho, 1989.

Ream, Lanny R. *The Gem & Mineral Collector's Guide to Idaho*. Gem Guides Book Co., Baldwin Park, California, 2000.

Sanborn, William B. *Handbook of Crystal and Mineral Collecting*. Gem Guides Book Co., Baldwin Park, California, 1987.

Sinkankas, John. *Gemstone and Mineral Data Book*. Van Nostrand Reinhold, New York, New York, 1981.

Sinkankas, John. *Field Collecting Gemstones and Minerals*. Geoscience Press, Phoenix, Arizona, 1988.

Stepanski, Scott, and Karenne Snow. *Gem Trails of Pennsylvania and New Jersey*. Gem Guides Book Co., Baldwin Park, California, 2000.

Voynick, Stephen M. *Colorado Rockhounding*. Mountain Press Publishing Company, Missoula, Montana, 1994.

Wilson, James R. *A Collector's Guide to Rock, Mineral & Fossil Localities of Utah*. Utah Geological Survey, Salt Lake City, Utah, 1995.

Zeitner, June Culp. *Midwest Gem, Fossil and Mineral Trails: Great Lakes States*. Gem Guides Book Co., Baldwin Park, California, 1999.

Zeitner, June Culp. *Midwest Gem, Fossil and Mineral Trails: Prairie States*. Gem Guides Book Co., Baldwin Park, California, 1998.

GOLD PROSPECTING

Basque, Garnet. *Gold Panner's Manual*. Heritage House Publishing Co., Ltd., Surry, British Columbia, 1991.

Basque, Garnet. *Methods of Placer Mining*. Sunfire Publications, Ltd., Langly, British Columbia, 1999.

Black, Jack. *Gold Prospector's Handbook*. Gem Guides Book Co., Baldwin Park, California, 1978.

de Lorenzo, Lois. *Gold Fever: The Art of Panning and Sluicing*. Gem Guides Book Co., Baldwin Park, California, 1978.

Garrett, Charles L. *Modern Metal Detectors*, Rev. ed. Ram Publishing Company, Dallas, Texas, 1995.

Klein, James and Jerry Keene. *How to Find Gold*. Keene Engineering Co., Northridge, California, 1996.

Klein, James. *Where to Find Gold in Northern California*. Gem Guides Book Co., Baldwin Park, California, 2000.

Klein, James. *Where to Find Gold in Southern California*. Gem Guides Book Co., Baldwin Park, California, 1994.

Lagal, Roy. *The New Gold Panning is Easy*. Ram Publishing Company, Dallas, Texas, 2001.

McCracken, Dave. *Gold Mining in the 21st Century*. Keene Industries, Northridge, California, 2001.

McPherson, Roger. *Modern Prospecting*. Gem Guides Co., Baldwin Park, California, 2002.

Ryan, A. H. *The Weekend Gold Miner*. Gem Guides Book Co., Baldwin Park, California, 1991.

Here are some websites of interest to rockhounds, treasure hunters and fossil buffs. You can't spend everyday searching for interesting collectibles and so for those down days you can search the Internet for virtual finds and interesting information. Enjoy the ride.

GOLD SITES

THE NEW '49ERS
http://www.goldgold.com/
- This recreational gold prospecting club helps hobby prospectors learn tips and techniques, provides opportunities for trips to their claims on the Klamath River in California and the creek near Happy Camp. The group also maintains claims in Arizona.

ORIGINAL SIXTEEN TO ONE MINE
http://www.origsix.com/
- This California gold mine is famous for the crystalline deposited in snowy white quartz. This commercial operation offers tours to the public and sells specimens and jewelry.

TOM ASHWORTH'S PROSPECTORS CACHE
http://www.tomashworth.com/
- This site has a plethora of information on gold mining, metal detecting and treasure hunting, including tips, how-to information, forums, chats and treasure hunting equipment for sale.

WORD WIDE GOLD NUGGETS
http://members.aol.com/worldwidegold/goldnuggets.htm
- Natural 24 karat gold nuggets from throughout the world are available to you through this site.

PROSPECTOR'S GOLD & GEMS
http://www.frii.com/~gold/gold.html
- This site sells gold nuggets, dust, platinum, emeralds and other precious things.

SHIPWRECK TREASURE SITES

MEL FISHER'S TREASURE SITE
http://www.melfisher.com/index1.html
- Explore Spanish treasure recovered from the coast of Florida by legendary treasure hunter Mel Fisher and his crew.

SPANISH MAIN TREASURE COMPANY
http://www.treasurenet.com/spanishmain/
- This salvage company, founded by Captain Carl Fismer in 1980, recovers artifacts and sunken treasure from shipwrecks around the world. They sell treasure, coins, artifacts and other items of interest to those who want to own a piece of history or a piece of eight.

EXPEDITION *WHYDAH*
http://www.whydah.com/
- The pirate ship *Whydah,* captained by infamous Sam Bellamy, sunk in 1717 off the coast of Cape Cod, taking one hundred forty-four men and booty from over fifty ships to the bottom. Treasure salvager Barry Clifford found the wreck and the salvaged artifacts can be seen online or at his museum in Providence town. The site has a ship's store.

INSTITUTE OF MARINE ARCHAEOLOGICAL CONSERVATION
http://www.imacdigest.com/
- Visitors can keep abreast of the current events and issues in treasure salvage here.

Quartz Crystals Sites

LEBOW CRYSTAL MINING CO.
http://www.lebowcrystals.com/
- Some of the best Arkansas quartz crystal clusters and points; Tabby crystal and double terminated points; and Brazilian crystal points are available for sale at this site.

SONNY STANLEY QUARTZ CRYSTALS
http://www.mtidachamber.com/stanley/index.html
- This Arkansas rock shop and museum sells all kinds of rocks and specializes in Arkansas quartz crystals.

STARFIRE QUARTZ CRYSTALS MINES
http://www.starfirecrystals.com/
- This is a crystal mine that sells all types of rock specimens and operates a crystal mine where you can dig your own crystals.

WEGNER QUARTZ CRYSTAL MINES
http://www.wegnercrystalmines.com/
- This is a crystal mine that sells all types of rock specimens and operates a crystal mine where you can dig your own crystals.

WORLD CHAMPIONSHIP CRYSTAL DIG ARKANSAS
http://www.mtidachamber.com/crystal-dig.htm
- Annual contest that offers prizes to the winner, crystal trophies and, of course, you keep all the crystals you dig.

QUARTZ, QUILTZ AND CRAFTZ FESTIVAL ARKANSAS
http://www.mtidachamber.com/crystal-festival.htm
- This is an annual crafts festival focusing on gems, minerals and oddly enough–quilts!

CRYSTAL HEAVEN MINING CO. INC.
http://www.crystalheaven.com/
- This site has much information on the digging, cleaning and sale of quartz crystals and metaphysics.

Rock Shops Sites

WRIGHT'S ROCK SHOP
http://www.wrightsrockshop.com/
- This Arkansas rock shop specializes in quartz crystals and clusters.

BOB'S ROCK SHOP
http://www.rockhounds.com/rockshop/table.html
- A non-commercial site that displays mineral specimens and provides topical information on locating gems and minerals.

ROCKHOUNDS INFORMATION PAGE
http://www.rahul.net/infodyn/rockhounds/rockhounds.htm
- This resource has a vast amount of information of interest to rockhounds.

OSOSOFT MINERAL CONNECTION
http://www.osomin.com/
- This site sells mineral specimens online, and has directories and website links to rock shops and mineral museums listed by state.

GEMS AND MINERAL SITES

CHRYSOPASE MINES OF AUSTRALIA
http://www.australianjade.com.au/index.html
- This Australian site tells you everything you want to know about chrysopase, including how to buy it.

GALLAGHER MINERALS
http://www.chrysocolla.com/
- Gallagher's sells arrowheads, artifacts, rough and specimen pieces along with cabochon stones.

FABRE MINERALS
http://www.fabre-minerals.com/
- This site sells high quality mineral specimens from around the world.

GEMSTONES CRYSTALS AND JEWELRY
http://www.mysticmerchant.com/
- Gem and jewelry enthusiasts will find gems, crystals, cabochons, faceted and one of a kind jewelry for sale at this site. Metaphysical books and information are also available.

TRINITY MINERAL COMPANY
http://www.trinityminerals.com/index.htm
- This shop has an interesting variety of gold and mineral specimens for sale.

THE ARKENSTONE
http://www.irocks.com/
- This California-based company supplies a variety of specimen rocks and minerals.

MINERAL MINER
http://www.mineralminers.com/
- This virtual gallery displays thousands of photographs of mineral specimens, crystals, gemstones, handcrafted jewelry, rough material and gift items.

HERKIMER DIAMONDS MINE
http://www.herkimerdiamond.com/
- These world-famous double-terminated quartz crystals are faceted by natural forces. The Herkimer Diamond Mine site includes information on the crystals, tips on how to mine, recent finds and specimens, jewelry and books for sale.

JOHN BETTS FINE MINERALS
http://www.johnbetts-fineminerals.com/
- New minerals are added every week to this site so customers will be pleased with the variety and constant arrival of new specimens.

THE MINERAL SHOWCASE
http://www.mineralshowcase.com/index.htm
- A plethora of minerals, fossils and rock are for sale to retailers, hobbyists and wholesalers.

METAL DETECTOR SITES

KELLYCO METAL DETECTORS SUPER STORE
http://www.kellycodetectors.com/indexmain.htmm
- This distributor supplies all brands of metal detectors and other tools for the metal detector hobbyists.

MINELAB, INC.
http://www.minelab.com.au/
- This Australian manufacturer created the broadband system of metal detectors and has many models from which to choose.

WHITES ELECTRONICS, INC.
http://www.whiteselectronics.com/
- Whites metal detectors are tried and true, a well-known brand.

TREASURE NET
http://www.treasurenet.com/
- This site keeps hobbyists up-to-date on what is being found, places to go and is a good addition to any metal detector's links.

FOSSIL SITES

COLLECTING FOSSILS IN CALIFORNIA
http://www.gtlsys.com/
- This site provides information on fossil sites in California and instructions on how to collect, prepare and display your finds.

AMBERICA WEST
http://www.ambericawest.com/index2.html
- Amber of all sorts is sold at this site. Many of the specimens have insects, plants, spiders and an occasional lizard encased in amber.

TWO GUYS FOSSILS
http://www.twoguysfossils.com/fish.htm
- Fossil fish from all over the world for sale.

STEVE'S FOSSIL SHARK TEETH
http://www.megalodonteeth.com/
- This site offers a wide variety of fossilized sharks teeth for sale including, great white teeth, megalodon teeth and prints, books and other items.

LAPIDARY AND GEMOLOGY SITES

JOHN MILLER'S GEMOLOGY AND LAPIDARY PAGES
http://www.tradeshop.com/gems/index.html
- This site has much information on faceting, as well as tips and techniques on gem cutting and lapidary arts. It is a comprehensive introduction to the jewelry trade.

MSI MULTISTONE INTERNATIONAL INC.
http://www.multistoneintl.com/rough/
- From meteors to incense burners, this place has it for sale.

GEMOLOGICAL INSTITUTE OF AMERICA
http://www.gia.org/
- This association is an educational institute who certifies gemologists and offers various correspondence courses in the area of gemology.

INTERNATIONAL COLORED GEMSTONE ASSOCIATION

http://www.gemstone.org/

- This association is a colored gemstone information resource, which includes gem lore, gems of the rich and famous, gemstone characteristics and more.

MAGAZINE SITES

LAPIDARY JOURNAL

http://www.lapidaryjournal.com/

- *The Lapidary Journal* is well-known trade magazine for gold and silversmiths, lapidary and gem cutters.

ROCK & GEM MAGAZINE

http://www.rockngem.com/

- *Rock & Gem Magazine* specializes in stories to fascinate gem trail readers and contains the latest lapidary news.

COLORED STONE MAGAZINE

http://www.colored-stone.com/

- *Colored Stone Magazine* informs readers of who, what, where and when gemstones surface around the world.

MINERAL LOCATOR INDEX

Photo Gallery

Quartz Crystal specimen

dredging

Hillside Prospecting in North Carolina

Checking the screen for Herkimer Diamonds

Staurolite (Fairy Stones) specimens

Photo Gallery

Screening sediment

Preparing to use the high-banker

Shark tooth collection

Using a rock pick to free a specimen

Searching for shark teeth in Florida

Time for a well-deserved break.

Gem Guides Publications

Rocks, Minerals, Gem Trails, & Fossils

Desert Gem Trails, _Strong._
  80 pgs., ISBN 0-910652-15-5, $5.00
Fee Mining and Rockhounding Adventures
  in the West, _Monaco._
  240 pgs., ISBN 1-889786-18-7, $12.95
Florida's Geological Treasures, _Comfort._
  160 pgs., ISBN 0-935182-95-0, $11.95
The Gem & Mineral Collector's Guide to Idaho,
  Ream. 80 pgs., ISBN 1-889786-13-6, $9.95
Gem Trails of Arizona, _Mitchell._
  224 pgs., ISBN 1-889786-14-4, $9.95
Gem Trails of Colorado, _Mitchell._
  144 pgs., ISBN 0-935182-91-8, $9.95
Gem Trails of Northern California, _Mitchell._
  160 pgs., ISBN 0-935182-67-5, $9.95
Gem Trails of Southern California, _Mitchell._
  216 pgs., ISBN 1-889786-25-X, $12.95
Gem Trails of Nevada, _Mitchell._
  192 pgs., ISBN 1-889786-15-2, $12.95
Gem Trails of New Mexico, _Mitchell._
  240 pgs., ISBN 1-889786-12-8, $12.95
Gem Trails of Oregon, _Mitchell._
  192 pgs., ISBN 0-935182-99-3, $10.95
Gem Trails of Pennsylvania & New Jersey,
  Stepanski and Snow.
  160 pgs., ISBN 1-889786-09-8, $12.95
Gem Trails of Texas, _Mitchell._
  160 pgs., ISBN 1-889786-11-X, $12.95
Gem Trails of Utah, _Mitchell._
  168 pgs., ISBN 0-935182-87-X, $9.95
Gems and Minerals of Washington, _Ream._
  217 pgs., ISBN 0-918499-09-7, $9.95
The GPS Guide to Western Gem Trails, _Kelty._
  240 pgs., ISBN 1-889786-20-9, $14.95
Midwest Gem, Fossil & Mineral Trails:
  Great Lakes States, _Zeitner._
  128 pgs., ISBN 1-889786-06-3, $10.95
Midwest Gem, Fossil & Mineral Trails:
  Prairie States, _Zeitner._
  128 pgs., ISBN 0-935182-94-2, $10.95
The Rockhound's Handbook, _Mitchell._
  194 pgs., ISBN 0-935182-90-X, $12.95
World of Gemstones, _Dud'a._
  Hard cover 92 pgs., ISBN 0-945005-30-X, $17.95

Beadwork Books

Beads: The Art of Stringing, _Ragan._
  58 pgs., ISBN 0-935182-44-6, $4.95

Jewelry Crafts

Contemporary Wire-Wrapped Jewelry, _Leonard._
  144 pgs., ISBN 0-935182-71-3, $15.95

Gembooks

Advanced Cabochon Cutting, _Cox._
  64 pgs., ISBN 0-910652-14-7, $5.00
The Art of Gem Cutting, _Dake._
  96 pgs., ISBN 0-935182-72-1, $5.50
Cabochon Cutting, _Cox._
  64 pgs., ISBN 0-935182-27-6, $5.00
Facet Cutter's Handbook, _Soukup._
  64 pgs., ISBN 0-910652-06-6, $5.00
Handbook of Crystal and Mineral Collecting,
  Sanborn. 82 pgs., ISBN 0-935182-48-9, $3.50
Handbook of Gemstone Carving, _Wertz._
  48 pgs., ISBN 0-935182-73-X, $4.50
Handbook of Lost Wax and Investment Casting,
  Sopcak. 64 pgs., ISBN 0-935182-28-4, $5.00
How to Create with Horseshoe Nails, _Link._
  22 pgs., ISBN 0-910652-26-0, $2.25
How to Design Jewelry, _Austin and Geisinger._
  30 pgs., ISBN 0-910652-28-7, $4.50
How to Make Wire Jewelry, _Jenkins and
  Thrasher._ 32 pgs., ISBN 0-910652-32-5, $5.00
How to Repair Jewelry, _Phelps._
  30 pgs., ISBN 0-910652-29-5, $5.00
How to Tumble Polish Gemstones, _Wexler._
  32 pgs., ISBN 0-935182-37-3, $5.00
How to Use Diamond Abrasives, _Riggle._
  30 pgs., ISBN 0-910652-30-9, $5.00
Jewelry Craft Made Easy, _French._
  72 pgs., ISBN 0-935182-60-8, $5.95
Jewelry Maker's Handbook, _Geisinger._
  72 pgs., ISBN 0-935182-61-6, $5.95
Jewelry Making for Beginners, _Soukup._
  48 pgs., ISBN 0-910652-17-1, $4.50

For the Younger Reader

Hello Arizona, _Tegeler._
  30 pgs., ISBN 0-943169-07-0, $3.50

Railroads

Trails Begin Where Rails End, _Manchester._
  Hard Cover, 160 pgs., ISBN 0-87046-081-1, $19.95

Cookbooks

Soup's On, _Hobbs._
  192 pgs., ISBN 0-9642012-1-6, $12.95

Hiking and Biking

50 Hikes in New Mexico, _Evans._
  239 pgs., ISBN 0-935182-77-2, $12.95
Day Hikes and Trail Rides in and Around
  Phoenix, _Freeman._
  304 pgs., ISBN 1-889786-10-1, $14.95

Travel and Local Interest

Arizona 101: An Irreverent Short Course for
New Arrivals, *Cook.*
80 pgs., 0-935182-80-2, $7.95

The Arizona Trivia Book, *Cook.*
192 pgs., ISBN 0-935182-51-9, $6.95

Arizona Roadside Discoveries, *Hutchins.*
160 gs., ISBN 1-889786-19-5, $12.95

Baby's Day Out is Southern California: Fun Places to
Go With Babies and Toddlers, *Holt.*
336 gs., ISBN 1-889786-26-8, $16.95

Destination: Phoenix, *Tegeler.*
244 pgs., ISBN 1-889786-01-2, $12.95

Discover Historic California, *Roberts.*
594 pgs., ISBN 1-889786-29-2, $15.95

Discover Historic Washington State, *Roberts.*
512 pgs., ISBN 1-889786-07-1, $13.95

Downtown L.A. : A Walking Guide, *Herman.*
280 pgs., 1-889786-17-9, $13.95

Dry Humor: Tales of Arizona Weather, *Cook.*
102 pgs., ISBN 0-935182-54-3, $6.95

Freeway Alternates: A Guide to Commuting in
Los Angeles and Orange Counties, *Rizzo.*
250 pgs., ISBN 0-935182-45-4, $10.95

Moving to Arizona, *Tegeler.*
188 pgs., ISBN 1-889786-02-0, $11.95

Retiring in Arizona, *Tegeler.*
176 pgs., ISBN 0-935182-89-6, $10.95

Saddleback Sightseeing in California, *Greenwald.*
208 pgs., ISBN 0-935182-58-6, $12.95

San Francisco Street Secrets, *Eames.*
170 pgs., ISBN 0-935182-75-6, $10.95

Historical & Old West

California Ghost Town Trails, *Leadabrand and
Broman.* 128 pgs., ISBN 0-935182-21-7, $7.95

Nevada Ghost Town Trails, *Broman.*
80 pgs., ISBN 0-935182-09-8, $7.95

Old West Trivia, *Bullis.*
160 pgs., ISBN 0-935182-62-4, $10.95

Prospecting & Treasure Hunting

Arizona's Golden Secret: How to Get Your Share
of Desert Gold!, *Wielgus.*
54 pgs., ISBN 0-9635601-0-7, $13.95

Dry Washing for Gold, *Klein.*
89 pgs., ISBN 0-935182-76-4, $7.95

Fee Mining and Rockhounding Adventures
in the West, *Monaco.*
240 pgs., ISBN 1-889786-18-7, $12.95

Gold Digger's Atlas, *Johnson.*
64 pgs., ISBN 1-889786-08-X, $6.00

Gold Fever: The Art of Panning and Sluicing,
de Lorenzo. 80 pgs., ISBN 0-935182-00-4, $6.95

Gold Prospector's Handbook, *Black.*
176 pgs., ISBN 0-935182-32-2, $10.95

Modern Prospecting, *McPherson.*
300 pgs., ISBN 1-889786-16-0, $14.95 (tent.)

Placer Gold Deposits of Arizona, *Johnson.*
103 pgs., ISBN 0-935182-33-0, $7.95

Placer Gold Deposits of Nevada, *Johnson.*
118 pgs., ISBN 0-89632-010-3, $7.95

Placer Gold Deposits of New Mexico, *Johnson.*
46 pgs., ISBN 0-89632-004-9, $5.95

Placer Deposits of the Sierra Nevada, *Morrison.*
160 pgs., ISBN 0-935182-97-7, $10.95

Placer Gold Deposits of Utah, *Johnson.*
26 pgs., ISBN 0-89632-007-3, $4.95

Recreational Gold Prospecting for
Fun and Profit, *Butler.*
206 pgs., ISBN 0-935182-98-5, $12.95

Sterling Legend: Story of the
Lost Dutchman Mine, *Conatser.*
94 pgs., ISBN 1-889786-23-3, $9.95

The Weekend Gold Miner, *Ryan.*
80 pgs., ISBN 0-935182-46-2, $5.50

Where to Find Gold & Gems in Nevada, *Klein.*
109 pgs., ISBN 0-935182-15-2, $8.95

Where to Find Gold in Northern California,
Klein. 125 pgs., ISBN 1-889786-05-5, $10.95

Where to Find Gold in Southern California,
Klein. 112 pgs., ISBN 0-935182-68-3, $7.95

Where to Find Gold in the Desert, *Klein.*
144 pgs., ISBN 0-935182-81-0, $7.95

Where to Find Gold in the Mother Lode, *Klein.*
129 pgs., ISBN 0-9620204-9-4, $9.95

HAVE WE MISSED A MINE OR COLLECTING SITE?

If you know of a mine or collecting site that should be included in future editions of this book, please bring it to our attention. Just fill out the bottom half of this sheet and return it to the address listed below. If your location is added to the next edition, we will send you a copy of the next edition free. Thank you for your help.

NAME OF MINE: _____

ADDRESS: _____

_____

_____

PHONE: _____

TYPE OF MATERIAL FOUND AT MINE: _____

_____

METHOD OF COLLECTING: _____
(EQUIPMENT USED) _____

COST: _____ **SEASON OF OPERATION:** _____

ADDITIONAL INFORMATION: _____

_____

_____

YOUR NAME & ADDRESS: _____

_____

_____

_____

RETURN TO: Attn: Editor
Gem Guides Book Co.
315 Cloverleaf Drive, Suite F
Baldwin Park, CA 91706